Wolfgang Bartknecht

Dust Explosions
Course, Prevention, Protection

With a Contribution from Günther Zwahlen
With a Preface by H. Brauer

Translation from German by
R. E. Bruderer, G. N. Kirby, and R. Siwek

With 295 Figures (some in Color)
and 26 Tables

Springer-Verlag Berlin Heidelberg New York
London Paris Tokyo Hong Kong

Dr. Wolfgang Bartknecht
c/o Ciba-Geigy A. G., Zentraler Sicherheitsdienst
CH-4002 Basel

Günther Zwahlen Dipl. Chem. HTL
Landskronstr. 12
CH-4143 Dornach

Translations:
R. E. Bruderer
c/o Ciba-Geigy Corporation, Toms River
N. J. 08753 USA

G. N. Kirby, Ph. D.
Ebasco Services Incorporated, New York
N. Y. 10048 USA

R. Siwek
c/o Ciba-Geigy AG, CH-4002 Basel

Title of the German Edition
Staubexplosionen. ISBN 3-540-16243-7
Springer-Verlag Berlin Heidelberg New York London Paris Tokyo

---

ISBN 3-540-50100-2 Springer-Verlag Berlin Heidelberg New York
ISBN 0-387-50100-2 Springer-Verlag New York Berlin Heidelberg

Library of Congress Cataloging-in-Publication Data
Bartknecht, Wolfgang. [Staubexplosionen. English] Dust-explosions: course, prevention, protection / Wolfgang Bartknecht; with a contribution from Günther Zwahlen; with a preface by H. Brauer; translation from the German by R.E. Bruderer, G.N. Kirby, and R. Siwek.
p. cm.
Translation of: Staubexplosionen. Bibliography: p.   Includes index.
ISBN 0-387-50100-2 (U.S.: alk. paper)
1. Dust explosions.   I. Zwahlen, Günther.   II. Title.
TH9446.D86B3713   1989   604.7--dc19   89-4187

This work is subject to copyright. All rights are reserved, whether the wole or part of materials is concerned, specifically the rights of translation, reprinting, reuse of illustrations, reciation, broadcasting, reproduction on microfilms or in other ways, and storage in data banks. Duplication of this publication or parts thereof is only permitted under the provisions of the German Copyright Law of September 9, 1965, in its version of June 24, 1985, and a copyright fee must always be paid. Violations fall under the procesucation act of the German Copyright Law.

© Springer Verlag Berlin Heidelberg 1989
Printed in Germany

The use of general descriptive names, registered names, trademarks, etc. in this publication does not imply, even in the absence of a specific statement, that such names are exempt from the relevant protective laws and regulations and therefore free for general use.

Product Liability: The publisher can give no guarantee for information about drug dosage and application thereof contained in this book. In every individual case the respective user must check its accuracy by consulting other pharmaceutical literature.

Typesetting, printing and book binding: Kieser, Augsburg
2152/3145-543210 – Printed on acid-free paper

Preface

Dust represents the most hazardous form of solid matter. It may result from natural or technical processes. Dust must be considered as a phase of its own with very specific characteristics. Dust is defined as fine particles of a solid dispersed in a gas. In considering explosions, the gas will initially be thought of as air with its usual oxygen content. In an explosion the dusty solid will always spontaneously oxidize.

The violence of an explosive oxidation increases with increased fineness of the individual solid particle because each particle's surface area increases quite rapidly in comparison with the total volume of the solid dispersed in the gas. An example may illustrate this statement. Consider a solid with an initial volume assumed to be 1 cm^3. If it exists in the form of a cube then its surface area will be 6 cm^2. However, if the same volume is visualized as particles with a 1 μ diameter, then the total surface area will be $6 \cdot 10^4$ cm^2. The surface of the dust is therefore ten thousand times larger than the reference cube. The actual effective conditions may even reach much larger ratios as with decreasing particle size the inner surface area which effects the oxidation process will generally be a multiple of the external surface area. It is the surface of the dust which makes the dust become such a hazardous material.

From a strictly scientific point of view, the course of an explosive oxidation process has so far only been investigated to a rudimentary degree. A satisfactory insight into dust explosions is therefore only possible from comprehensive experimental tests. Due to the high cost of suchtests, it is necessary to analyze the results with great care and to summarize the data with the help of empirical characteristic parameters. These empirical parameters are the basis for the guidelines which have been developed in order to help prevent explosions or at least to limit the consequences.

The author has summarized today's knowledge of the cause, course and consequences of dust explosions in an outstanding fashion. Explosions became the problem around which his professional life developed with great success. Whoever is confronted with dusts, be it the field engineer or the scientist, will benefit from this book and increase the safety of the facility for which he is responsible.

Heinz Brauer
Full Professor of Reaction Technology
Technical University Berlin

Table of Contents

1	*Introduction*	1
2	*Historical Review*	2
2.1	Occurrence of Dust Explosions	2
2.2	The Nature of Dust Explosions	10
2.3	Apparatus for the Testing of Airborne Dusts	15
3	*Dust as a Dispersed Substance*	24
4	*Material Safety Specifications*	27
4.1	Preliminary Remarks	27
4.2	Material Safety Specifications of Dust Layers (G. Zwahlen)	28
4.2.1	Flammability	28
4.2.2	Burning Behavior	29
4.2.2.1	Combustibility Test at Room Temperature	29
4.2.2.2	Combustibility Test at Elevated Temperature	30
4.2.2.3	Burning Rate Test	31
4.2.3	Deflagration	32
4.2.3.1	Screening Test for Deflagration	32
4.2.3.2	Laboratory Test for Deflagration	33
4.2.4	Smolder Temperature	34
4.2.4.1	Determination of the Smolder Temperature	34
4.2.5	Autoignition	36
4.2.5.1	Determination of the Relative Autoignition Temperature, as per Grewer	37
4.2.5.2	Hot Storage Test in the Wire Mesh Basket	39
4.2.6	Exothermic Decomposition	40
4.2.6.1	Determination of the Exothermic Decomposition Temperature in an Open Vessel, as per Lütolf	40
4.2.6.2	Determination of an Exothermic Decomposition in an Oven Purged with Nitrogen, as per Grewer	44
4.2.6.3	Differential Thermal Analysis	44
4.2.6.4	Determination of an Exothermic Decomposition Under Choked Heat Flow	44
4.2.7	Explosibility	47
4.2.7.1	Impact Sensitivity	48

4.2.7.2	Friction Sensitivity	49
4.2.7.3	Thermal Sensitivity	50
4.3	Material Safety Specifications for Dust Clouds Describing the Explosion Behavior	51
4.3.1	Combustible Dusts	51
4.3.1.1	Preliminary Remarks	51
4.3.1.2	Particle Size Distribution	52
4.3.1.3	Explosibility	53
4.3.1.4	Explosible Limits	54
4.3.1.5	Explosion Pressure Versus Explosion Violence	56
4.3.2	Flock	81
4.3.2.1	Preliminary Remarks	81
4.3.2.2	Explosible Limits	82
4.3.2.3	Explosion Pressure / Violence of Explosion	84
4.3.3	Hybrid Mixtures	86
4.3.3.1	Preliminary Remarks	86
4.3.3.2	Explosible Limits	87
4.3.3.3	Explosion Pressure / Violence of Explosion	88
4.3.4	Conclusions	92
4.4	Safety Characteristics of Airborne Dust Describing the Ignition Behavior	93
4.4.1	Minimum Ignition Energy	93
4.4.1.1	Preliminary Remarks	93
4.4.1.2	Apparatus for the Determination of the Minimum Ignition Energy	93
4.4.1.3	Ignition Behavior of Combustible Dusts	96
4.4.1.4	Ignition Behavior of Flock	109
4.4.1.5	Ignition Behavior of Hybrid Mixtures	112
4.4.1.6	Conclusions	114
4.4.2	Ignition Temperature	115
4.4.2.1	Preliminary Remarks	115
4.4.2.2	Apparatus for Temperature Determination	116
4.4.2.3	Ignition Effectiveness of a Glowing Coil	117
4.4.2.4	Conclusions	118
4.5	Safety Characteristics of Airborne Dusts Describing the Course of an Explosion in Pipelines	119
5	*Protective Measures Against the Occurrence and Effects of Dust Explosions*	125
5.1	Preliminary Remarks	125
5.2.	Preventive Explosion Protection	127
5.2.1	Preliminary Remarks	127
5.2.2	Prevention of Explosible Dust/Air Mixtures	128
5.2.3	Prevention of Dust Explosions by Using Inert Matter	129
5.2.3.1	Admixture of Nitrogen	129
5.2.3.1.1	Preliminary Remarks	129

5.2.3.1.2	Combustible Dusts	130
5.2.3.1.3	Hybrid Mixtures	137
5.2.3.1.4	Use of Vacuum	140
5.2.3.1.5	Admixture of Solids	141
5.2.4	Prevention of Effective Ignition Sources	144
5.2.4.1	Preliminary Remarks	144
5.2.4.2	Mechanically Generated Sparks	145
5.2.5	Hot Surfaces / Autoignition	154
5.2.6	Static Electricity	156
5.2.7	Conclusions	160
5.3	Explosion Protection Through Design Measures	161
5.3.1	Preliminary Remarks	161
5.3.2	Explosion Pressure-resistant Design for the Maximum Explosion Pressure	163
5.3.2.1	Explosion Pressure-resistant Design	163
5.3.2.2	Explosion Pressure Shock-resistant Design	164
5.3.3	Explosion Pressure-resistant Design for a Reduced Maximum Explosion Pressure in Conjunction with Explosion Pressure Venting	166
5.3.3.1	Preliminary Remarks	166
5.3.3.2	Explosion Pressure Venting of Vessels	167
5.3.3.3	Explosion Pressure Venting of Elongated Vessels (Silos)	187
5.3.3.4	Explosion Pressure Venting of Pipelines	200
5.3.4	Explosion-resistant Construction for Reduced Maximum Explosion Pressure in Conjunction with Explosion Suppression	203
5.3.5	Technical Diversion or Arresting of Explosions	215
5.3.5.1	Preliminary Remarks	215
5.3.5.2	Extinguishing Barrier	215
5.3.5.3	Rotary Air Locks (Rotary Valves)	229
5.3.5.4	Rapid-Action Valves: Gate or Butterfly Type	232
5.3.5.5	Rapid-Action Valve: Float Type	241
5.3.5.6	Explosion Diverter	243
5.3.6	Conclusions	245
6	*Concluding Remarks*	246
7	*Acknowledgements*	247
8	*Appendix*	248
8.1	Explosion Pressure Venting	248
8.1.1	Vessel: Area Determination by Calculation or Nomogram	248
8.1.2	Elongated Vessels (Silos)	249

9	*References*	254
10	*Symbols and Abbreviations*	260
11	*Conversion Factors*	263
12	*Subject Index*	265

1 Introduction

In past years, the causes of dust explosions have been systematically researched by laboratory-testing the combustion, ignition, and explosion behavior of combustible dust. Step by step, the results and the experience gained have been transferred to industrial practice. Yet, dust explosions still occur, causing major damage and sometimes fatalities. The "Berufsgenossenschaftliche Institut für Arbeitssicherheit" in Bonn [1] lists 357 dust explosions which occurred from 1976–1980 in the Federal Republic of Germany, but only 155 of these cases had been officially reported. In contrast, the "Verband der Sachversicherer" (VDS) (organization of insurance carriers) states 300 cases per year with damage exceeding 50,000 DM each.

In conjunction with the above-mentioned laboratory tests, large, practice-related, full-scale tests were carried out with combustible dusts. The purpose was to develop preventive measures and to test design measures which may not avert an explosion but should limit its dangerous results to an acceptable level. The experience gained over the past years by applying such measures in facilities which process and convey combustible dusts indicates that damage to equipment and personnel can definitely be prevented.

The aim of this publication is to assist people dealing with industrial applications. It reflects the present state of the art and current knowledge of the course of a dust explosion, including the inherent danger of a dust layer.

2 Historical Review

2.1 Occurrence of Dust Explosions

Dust explosions have been known for approximately 200 years, ever since the wind mill was introduced in 1752–1756 for the purpose of grinding cereal grains.

The first explosion which was recognized as a dust explosion occurred in Italy on December 14, 1785. It was reported by the Turin Academy of Science as a flour dust explosion in a warehouse in Turin.

Five additional explosions creating considerable excitement occurred over the next 100 years.

Table 1. Dust explosions

Year	Location	Installation	Dust type	Damage
1785	Turin (Italy)	Warehouse	Flour	Warehouse destroyed
1858	Stettin (Poland)	Roller mill	Grain	Mill building destroyed
1860	Milwaukee (USA)	Mill	Flour	Mill building destroyed
1864	Mascoutah (USA)	Mill	Flour	Mill building destroyed
1869	unknown (Germany)	Mill	Pea flour	Local damage to mill
1887	Hameln (Germany)	Silo	Grain	Silo and building destroyed

In 1887, a grain dust explosion destroyed a silo of the "new Wesermühle" in Hameln, Germany (Table 1). The journal of the "Verein Deutscher Ingenieure" of November 1887 described the explosion as follows: "This accident is unique on the continent. Until now, we had no idea of the enormous destructive effects of such forces."

With increasing industrialization and the change from smaller facilities to large industrial complexes, the frequency of dust explosions has increased since 1887.

2.1 Occurrence of Dust Explosions

Fig. 1 a/b. The "Neue Wesermühle" in Hameln, Germany after the explosion in 1887

Most early dust explosions occurred in places where production and dust generation were high due to size and productivity. Up to 1922, the USA and Canada experienced 217 dust explosions. These involved organic dusts from mills, elevators, and silos, starch plants and refineries, as well as plants processing aluminum, chocolate, paper, rubber, seasoning, etc. The multitude of installations affected by dust explosions is striking.

A more recent USA statistic for the time period 1900–1952 lists the dust types involved in 769 explosions (Table 2). The total damage amounted to 88 million dollars and involved 464 casualties and 1229 injured.

Table 2. Type of dusts involved in dust explosions in the USA (1900–1952)

%	Number of explosions	Type of dust
24.8	191	Grain
16.8	129	Wood
14.7	113	Feedstuff
13.1	101	Flour
5.6	43	Starch
4.8	37	Cork
3.4	26	Sugar
3.3	25	Plastics
3.1	24	Sulfur
3.1	24	Malt
1.8	14	Bark
5.5	42	Miscellaneous
100	769	

But not only the industrial establishment experienced dust explosions. The Pacific Northwest (USA) registered 300 dust explosions in combines during harvest with damages exceeding 1 million dollars.

Statistical information on dust explosions in Germany is scarce for that time period. Nevertheless, 66 sugar dust explosions were reported for the time period 1890–1922, with moderate to catastrophic results. One of the worst explosions occurred on March 16, 1917 in Frankenthal, killing 6 operators and causing considerable damage (Fig. 2).

Other branches of industry experienced dust explosions during the same time period. 77 incidents were registered involving coal dust, dyestuffs, soot, and aluminum. Coal dust explosions alone left 404 dead and 275 injured.

Approximately 30 years ago, one of the most comprehensive books on dust explosions was published by W. H. Geck [3]. The author, who lost his own business due to a wood dust explosion with subsequent fire and who survived a severe sugar dust explosion, described 116 dust explosions in various major industries. The probable causes and effects are analyzed based on his own observations and over 30 years of experience as a consultant (Fig. 3).

In his work he reaches the remarkable and still valid conclusion that every dust may behave differently in different plants due to the influence of innumerable factors unique to each industrial environment.

The years 1976–1978 saw numerous grain dust explosions in silos in the USA. The incidents caused extensive damage and resulted in numerous fatalities and in-

Fig. 2. Effects of a sugar dust explosion [3]

Fig. 3. Wood dust explosion in a furniture factory with subsequent fire [3]

juries. Many symposia dealt with and are still dealing with the causes of these dust explosions. The discussions have concentrated on the effectiveness of preventive measures and other design measures to reduce the consequences of such mishaps to a minimum.

Nowadays, dust explosions are – in the true sense of the word – commonplace in the Federal Republic of Germany, although only a fraction are recorded, as men-

Fig. 4. "Bremer Roland Mill" after a flour dust explosion, 1979

tioned earlier [1]. But not all the mishaps have such a terrible outcome as the 1979 flour dust explosion in the "Bremer Roland Mill" (Fig. 4) which left 14 dead, 17 injured, and property damage of 100 million DM.

357 dust explosions have been analyzed and the frequency of involvement of the various dust categories tabulated (Fig. 5).

Almost one-third of the explosions were caused by wood dusts and every fourth explosion happened in the food and feedstuff industry. These figures are practically identical with the ones recorded in the USA for the time period 1900–1952 (Table 2).

In Germany, as well as in the USA, most of the dust explosions occurred in the industries cited above. The percentages given in Fig. 5 are therefore representative of both countries.

Figure 6 shows the percentage of equipment categories involved in the dust explosions. Every fifth of these happened in a silo or bunker. Such silo explosions are often spectacular, especially if an entire cluster of silos is involved (Fig. 7). Grinding and conveying systems, as well as dust collectors, participate at almost the same ratio.

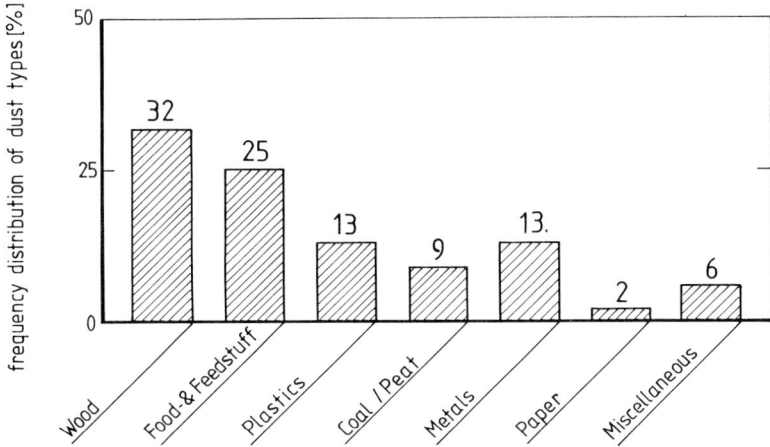

Fig. 5. Frequency distribution of dust types involved in 357 dust explosions [1] (1965–1980)

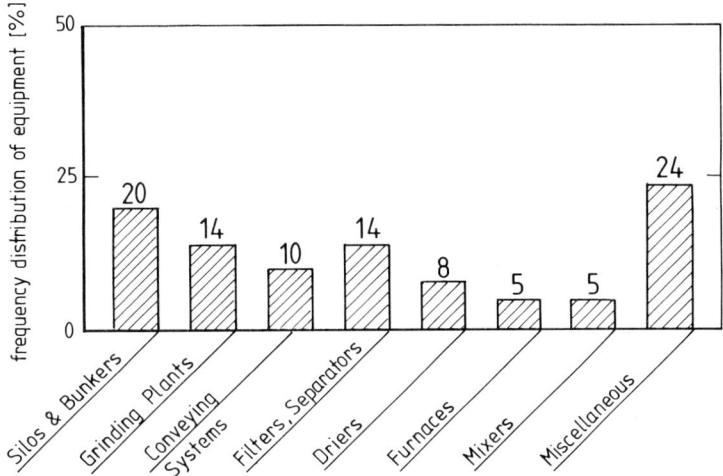

Fig. 6. Frequency distribution of types of equipment involved in 357 dust explosions [1] (1965–1980)

In conclusion, Fig. 8 represents the frequency distribution of the various ignition sources responsible for the above dust explosions. Although it is sometimes very difficult to determine the actual ignition source, it is apparent that mechanically produced sparks present the most frequent source in industrial practice, with 29%. This includes sparks generated through friction, grinding, and impact. Therefore, it is not surprising that present research is concentrated on the effectiveness

Fig. 7. Effects of a grain dust explosion in a group of silos

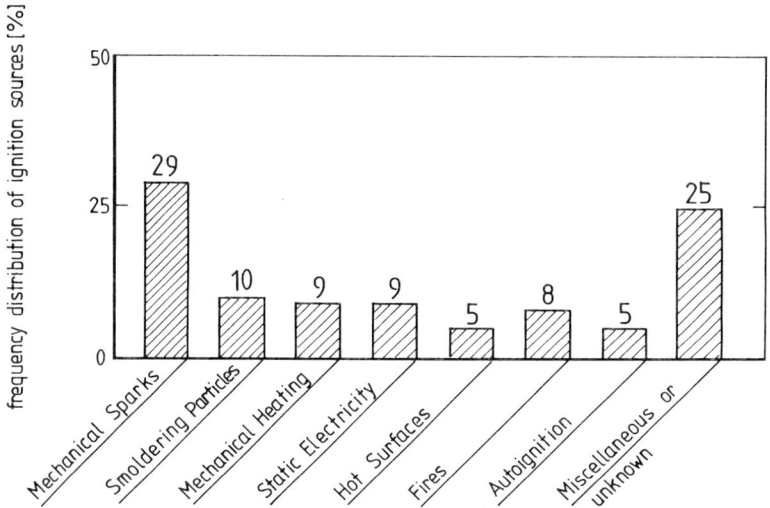

Fig. 8. Frequency distribution of ignition sources responsible for 357 dust explosions (1965–1980)

of such sources in the case of dust/air mixtures. All other sources combined participate practically with the same probability as mechanical sparks. The outlined case-histories, which are certainly not complete, may serve as an overview of the dangers which have been known to exist for the past 200 years in processing or handling combustible dusts. The often-voiced opinion that a plant which processes dusts

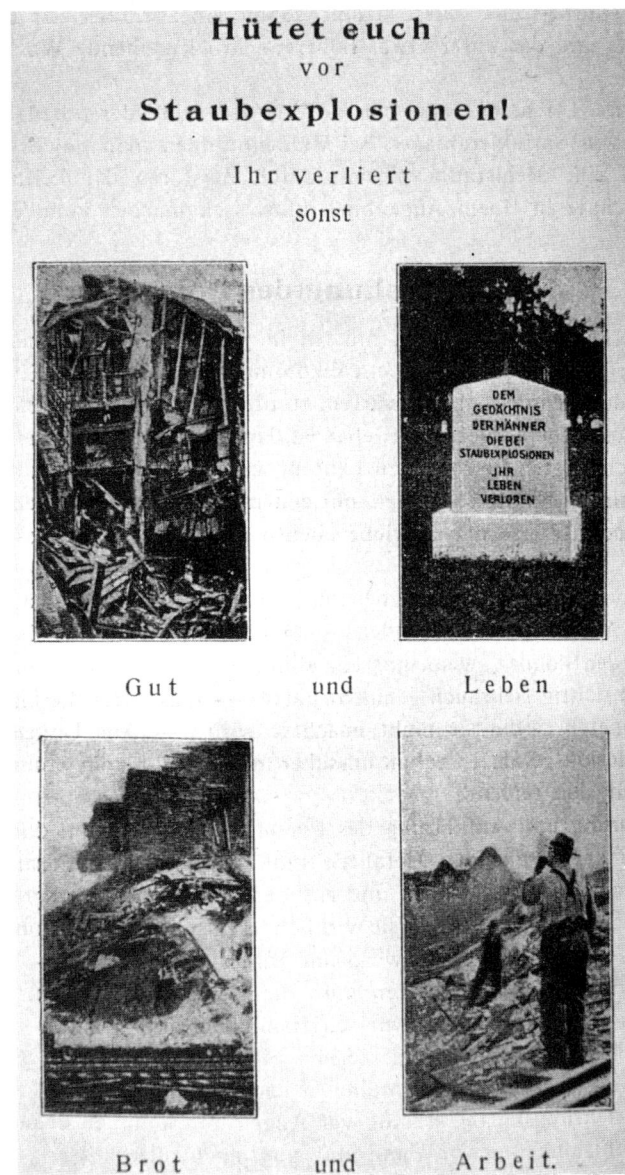

Fig. 9. Educational poster based on a design of US Departments of Agriculture and Interior

is safe because it has not experienced an explosion in the past years or decades has been frequently disproven in practice.

The fact that about 36% of all explosions have been caused by human error, i.e., thoughtlessness, negligence, indifference, and ignorance shows clearly that effective instructions and training of employees are essential.

In this context, the poster designed by the US Departments of Agriculture and Interior at the turn of the century is still valid with its slogan: "Watch out for dust explosions. You may lose property and life, bread and work".

2.2 The Nature of Dust Explosions

Originally it was thought that organic dusts were incapable of exploding. However, a publication [4] from 1878 states that bakers got rid of flies and moths in their bakeries by throwing very dry flour into the air and igniting the flour dust cloud. One wonders about the often frightening experience of blown-out bakery windows.

Because dust explosions as such could not be explained, a flammable gas was usually sought as the cause. After a dust explosion in the roller mill in Stettin in 1858, one group thought that alcoholic vapors generated from fermenting paste had ignited in a vent duct. Others felt that decomposing paste generated swamp gas, which was ignited by encroaching sparks from a nearby fire.

Also, an explosion in 1890 in a brown coal manufacturing plant was attributed to exploding gases resulting from oven-drying of brown coal.

Even in 1917, it was felt that the ignition of explosive sewer gases, which were next to the location of the mishap, caused a severe sugar dust explosion.

The first to recognize that organic dusts are explosible was M. Faraday. He proved that the severity of the coal mine explosion in Haswell in September 1844 was due to the presence of coal dust which was dispersed through the primary firedamp (methane) explosion. These findings were later confirmed by others: Verpilleux (1867), Vital (1875), Galloway (1881), Hilt and Marggraff (1884) as well as Treptow (1888) [5]. In 1881, in England, Galloway [6] wrote that certain coal dusts, when dispersed in pure air, would propagate a flame over an indefinite distance. The initial ignition was due to either an explosive charge or a methane explosion, which also dispersed the dust.

At the same time, French researchers had a different theory. They stated: "In our opinion we have proof that coal dust does not present a real danger in the absence of methane. The dust plays only a major roll in so far as it increases the danger of a wavetype explosion." Since the followers of the English and French points of view were opposing each other in the "Prussian firedamp commission", no consensus could be reached. In 1822, the Germans decided to start their own testing program to determine the danger of coal dust. Therefore, all experimental test programs at that time were aimed at answering the question of whether organic and metallic dusts were explosible or not. The pertinent discoveries are shown in Table 3.

Holtzwart and Meyer showed that fermentation gases generated by the drying of brown coal are not flammable and therefore not responsible for dust explosions. Dust from brown coal is explosible by itself. In 1885, Engler made some very significant discoveries: He experimented with soot and charcoal dust mixed in methane or lighting gas.

The components were non-explosible by themselves but the so-called hybrid mixture did explode.

Stockmeier thought that this effect influenced the ignition behavior of aluminum bronze. He proved that the inherent hygroscopic moisture did form hydrogen-air mixtures which in turn formed a hybrid mixture with the aluminum bronze.

2.2 The Nature of Dust Explosions

Table 3. Discoveries concerning the explosibility of combustible dusts

Year	Name	Explosible dust
1844	M. Faraday	*Coal dust*
1878	R. Weber	*Flour*
1885	Engler	Mixtures of *flammable gases* and *dust*
1891	R. Holtzwart and E. v. Meyer	*Brown coal dust*
1891	R. Holtzwart and E. v. Meyer	*Dust clouds* ignited with electrical sparks
1899	H. Stockmeier	*Aluminum dust* ignited with electrical sparks

Barely 100 years later, G. Pellmont [7] reported extensively on the ignition and explosion behavior of such hybrid mixtures.

As early as the turn of the century, it was noted that organic and metallic dusts, once dispersed in industrial applications, have to be considered explosible, whatever the "conditions for ignition". Weber and Stockmeier detected as early as 1878 the importance of particle size in assessing the danger potential of an explosion. They noticed that the flour which was most readily dispersed reacted most violently. Also Weinman, who carried out tests at the Bergfiskalischen test site in 1922 in Neunkirchen (Saar, Germany) concluded that the finer the dust, the easier the initiation of an explosion.

In 1925, the effect of particle size was deduced by Beyersdorfer [4] in the following way: Starting with a 10-cm cube, the volume would be $V = 0.001$-m^3 and the total surface area $SA = 0.06$-m^2. Then the 3 coordinates were subdivided by decimal fractions and the sum of the surfaces of the thus-generated cubes compared with the new edge (Fig. 10).

The increase of the total surface area with decreasing cube edge for the particle becomes substantial. Every surface has the tendency to adsorb the surrounding gaseous media; therefore, very fine dusts, which have a high surface activity, must react more violently than coarse dusts.

At the same time, Jaeckel proved experimentally the validity of the claim. In grinding plate sugar in a small air-tight ball mill, a distinct negative pressure was recorded after 24 h. This was explained by the adsorption of air on the newly created large surfaces.

Dusts therefore have a marked surface activity. Particle size and the resulting explosion hazard are approximately inversely proportional. Beyersdorfer was one of the first to recognize the danger of standing dust layers which cannot be the immediate cause for an explosion. However, if such a layer is dispersed because of a primary dust explosion, the ensuing secondary dust explosion may be devastating. Beyersdorfer calculated the maximum allowable dust accumulation on the 6 surfaces of a cubical work area so that the dispersed dust would be just below the explo-

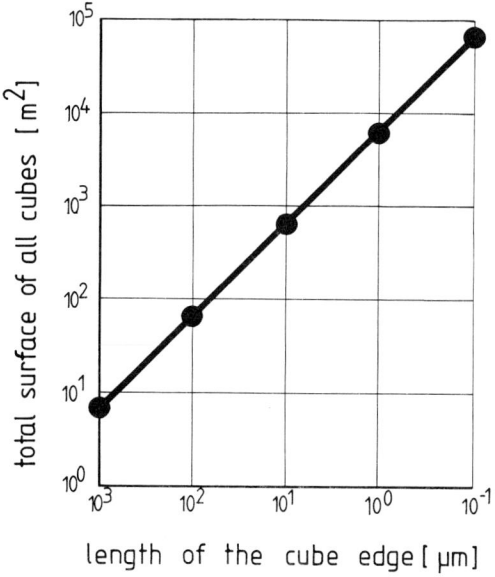

Fig. 10. Surface increase from subdividing a large cube into smaller cubes (length of original cube edge 10-cm)

sible threshold. He based this on sugar dust with a lower explosible limit of 21 g/m³ (Fig. 11).

The smaller the work space, the smaller the tolerated dust accumulation presenting no dust explosion hazard. Dust accumulation of a few hundredths of a mm suffice to create an explosible mixture. Cubically-shaped work spaces pose the least hazard. The following equation relates work space parameters, the properties of the dust, and the acceptable height or thickness of a dust accumulation:

$$\Delta = \frac{\frac{C}{2S}}{\frac{1}{l}+\frac{1}{w}+\frac{1}{h}}$$

where
Δ = acceptable height or thickness
C = lower explosible limit
S = density
l = length
w = width
h = height

Analyzing this equation for the acceptable height Δ of a dust accumulation gives the following:

a) the acceptable height Δ is smaller for non-cubical work spaces than for cubical ones and
b) it is also smaller for a narrow building with low ceilings than for wide and high spaces.

Beyersdorfer was also the first to make suppositions with regard to the mechanics of a dust explosion. He assumed that the explosion of an organic dust was in reality a gas explosion. He supported this with the following observation. A non-ex-

2.2 The Nature of Dust Explosions

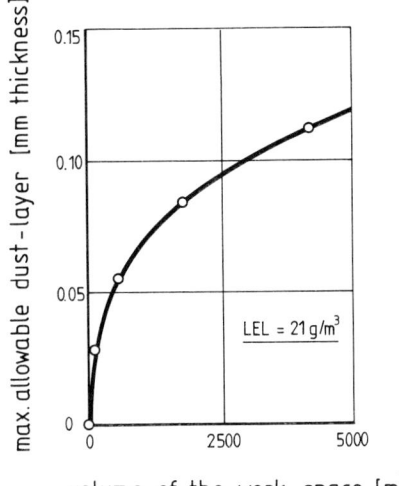

Fig. 11. Accumulations of sugar dusts in a cubical work space which will not present an explosion hazard if swirled up

plosible amount of sugar dust dispersed into a heated space would rapidly lower the temperature 10–15 °C. Afterwards the temperature would rise a few degrees above the initial temperature. He assumed that the observed temperature increase was due to a chemical reaction between "sugar gas" and air. Such a heat generation was thought to be the result of an oxidation, as atmospheric nitrogen does not qualify as a reactant. Therefore, the process of ignition has to occur in two phases:

1 "gasification of the sugar"
2 "oxidation of the sugar gas"

The released heat from the oxidation will ignite the sugar dust/air mixture if there is ample oxygen present and the temperature is high enough. As a further proof that organic dust explosions are secondary phase gas explosions, Beyersdorfer mentions the fact that only combustible gases will burn with a flame. Not the solid or liquid phase of stearin of a candle burns, but rather the stearin vapor.

The above-mentioned model for the start and the progression of a reaction is certainly not generally valid, especially in conjunction with metallic dusts [8, 9, 10]. The products of combustion of an aluminum dust explosion are similar in form and size to the single particles of the burnt dust.

This could hardly be the case if an evaporation had preceded the combustion. In addition, the evaporative temperature of aluminum is so high that the formation of metallic vapor is rather unlikely at the time of ignition. This is even more valid for zirconium dust, which has a substantially higher evaporative temperature. Explosions with aluminum dusts result in higher pressures than explosions with normal organic dusts. From this observation, it can be concluded that the pressure development is solely dependent upon the temperature of combustion, i.e., expansion of the residual gases, particularly nitrogen.

For the observed metallic dust explosions, radiation is very important for the transport of liberated energy. Gliwitzky and recently Leuschke [11] have shown experimentally that swirled-up metallic dust in air will explode through intensive light radiation alone without conduction and convection.

Certain types of organic dusts with low minimum ignition energy and autoignition temperature can be ignited either by the radiation of a flash bulb or from a dust explosion proper [11]. In the test, radiation seemed to be solely responsible because a glass window separated the fuel from the ignition source. Therefore, it can be considered proven that the radiation from dust explosions participates to a great extent in the energy transfer.

Zehr [12] recently confirmed the basic idea of Beyersdorfer with regard to the combustion mechanics of combustible dusts. He observed that the combustion of solid particles consisted of a reaction of oxygen in the gas phase surrounding the particle surface. The combustible matter reacts to produce flames, i.e., the previously evaporated or smoldering gases oxidize or react in the flame with the diffused oxygen. After ignition, either by an outside source or by autoignition – provided the conditions are right – the course of the incipient combustion of the dust/air mixture will liberate energy, which is transported mainly through radiation to the surrounding volume containing unreacted dust particles. Convection contributes a minor share.

The combustion of 10 g of product will result in the following:

Substance	Temp.increase
Sulfur	135°C
Magnesium	240°C
Aluminum	340°C

The reaction propagates three dimensionally at a relatively high velocity. The reactivity of the dust/air mixture varies with the type of dust. The analysis of coal dust explosions has consolidated these findings [6]. Initially, it was thought that the explosion occurred mainly in the gas phase. The volatile fraction of the ignited coal particles was claimed to support the reaction. Others suggested that the oxidation process at the surface of the coal particles was the determining factor. In recent years, however, it is thought that the expelled volatile components, as well as the solid coal substance determine the course of the explosion.

The volatile fraction seems to play a major role only at the start of the explosion. This seems to be supported by the results from experiments made worldwide.

The predominant interpretation describes the combustion of the coal particle in 3 stages:

– heat-up of the granule and pyrolysis (formation and liberation of the volatile fraction and tarry products)
– ignition and combustion of pyrolytic products
– ignition and combustion of the coke- and coal residues.

The time sequence of the appearance of the three phases depends mainly upon the particle size and the heat rate. Howard and Essenhigh [13, 14] found with particles in the mm range and a heat rate of approximately 10^4 °C/min a distinct dividing line between the liberation of volatiles and combustion. With smaller particles and a higher heat rate, however, the surface of the granule caught fire before the volatiles were released.

The supposition that a coal dust explosion occurs in three stages may be proven by the fact that the heat rate is estimated to be in the $10^6 - 10^9$ °C/min range and the particles which sustain the reaction are relatively fine.

2.3 Apparatus for the Testing of Airborne Dusts

The attempt to simulate dust explosions in available laboratory equipment and to record the course of the explosion in a reproducible fashion was met with substantial difficulties. The prerequisite for such tests was the existance of a uniform and reproducible dust/air mixture. Such a requirement seemed easiest achieved in small equipment for the determination of the most important explosion characteristics and the assessment of the dangers of an explosion. The equipment was mostly made of glass, which allowed a visual check of the homogeneity of the dust mixture.

Vital [15] was one of the first to conduct comparative tests in 1875 to determine the hazards of various dust types. He blew a dust cloud through a gas flame into a long pipe which had a ball made of elder wood at the end. From the shape and length of the explosion flame as well as the throw of the wooden ball, Vital estimated the "relative hazard" of the tested dust.

Approximately at the same time, R. Weber ran flour tests using the equipment shown in Fig. 12. The dust was deposited on top of a sieve. Through shaking action, the product fell into a vessel containing an ignition source (not shown in Fig. 12).

By changing the revolutions of the cogwheel, by adjusting the stroke of the sieve and by combining sieves with varying mesh sizes, dust streams of differing densities could be produced. The stream was also reduced to just that volume that

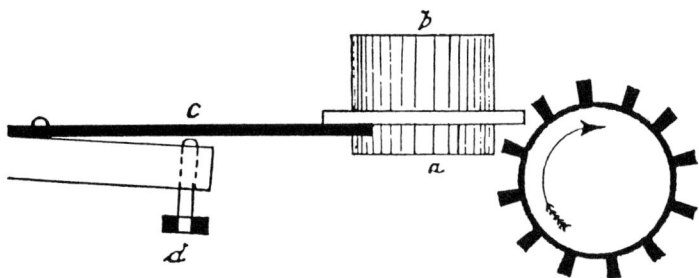

Fig. 12. Apparatus for the determination of the lower explosible limit, as per Weber

barely gave ignition. In such fashion, Weber determined the lower explosible limit of flour at 25 g/m³.

In 1911, Tifanel and Dürr used a photographic method to determine the relative danger of a few dust types. A pre-weighed amount of dust was conveyed by a stream of oxygen into a well-defined gasoline flame and a picture taken of the length and size of the flame. Mainly flour, sugar, and Lycopodium were tested.

Figure 13 shows the apparatus R. Bauer developed in 1917 for the determination of the lower explosible concentration of aluminium dust. The apparatus consisted of a vertical tube with an impeller with electric drive in the lower portion, plus an ignition source and a loose cover. After the addition of a known amount of dust, the motor was started and the ignition activated. With this arrangement, the lower explosible limit of aluminium dust was determined at approximately 400 g/m³.

The apparatus which was developed by Steinbrecher [6] for the determination of the lower explosible limit of industrial dusts (Fig. 14) differs from the previously described unit in its singly activated dust dispersion system through a blast of air. The combustion chamber consisted of a vertical glass pipe with a hemispherical end and a volume of 0.135 l. The pipe also housed an ignition source and a thermocouple.

The dust sample was stored just ahead of the explosion chamber and then dispersed in the cylinder by a jet of air. These igniton tests often destroyed the glass tube.

Between 1933 and 1935, the Chemisch-Technische Reichsanstalt [8] (German Chemical Technical Institute) used the apparatus shown in Fig. 15 for the determination of the explosibility of dust/air mixtures.

The dust was first deposited at the bottom of the apparatus and then swirled up by a jet of compressed air discharged through a nozzle. A glowing wire was used as an ignition source. The test results indicated that zirconium was especially explosible over a markedly wide range of concentrations.

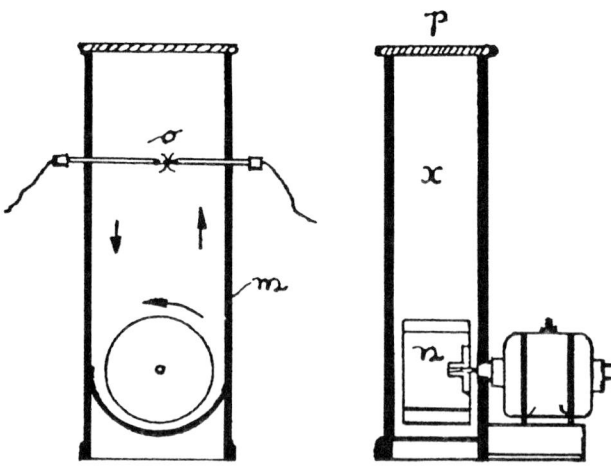

Fig. 13. Apparatus for the determination of the lower explosible limit, as per R. Bauer

2.3 Apparatus for the Testing of Airborne Dusts

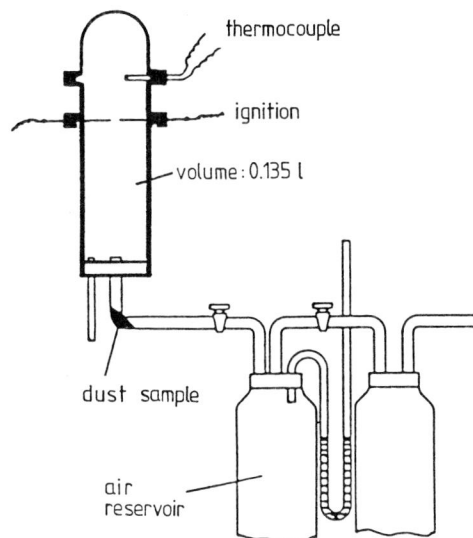

Fig. 14. Apparatus for the determination of the lower explosible limit, as per Steinbrecher

Until recently, Eckhoff [17] used an oxyacetylene torch for ignition in testing the explosibility of dust in order to eliminate the risk of a misjudgement.

The hazard of a combustible dust is not only defined by its explosibility and explosible range but also by the pressure and violence of its combustion. In such a context, the maximum explosion pressure and the pertinent pressure rise are of spe-

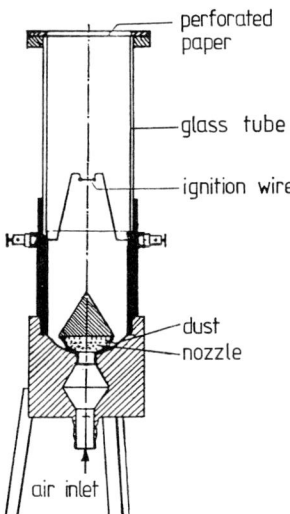

Fig. 15. Apparatus used by the Chemisch-Technische Reichsanstalt to investigate the explosibility of dust/air mixtures

18   2 Historical Review

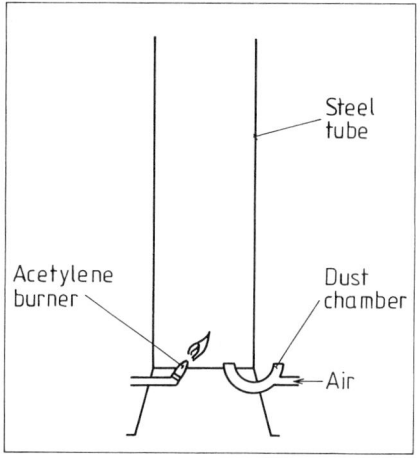

Fig. 16. Apparatus for the determination of the explosibility of dust/air mixtures, as per Eckhoff (D = 14-cm, H = 40-cm)

cial interest. Closed equipment is needed to test for these parameters. Obviously, the test equipment shown in Fig. 12–16 is unsuited for this purpose.

In 1937, Mason and Taylor [18] used a 0.75-l cylindrical glass vessel as the explosion chamber (Fig. 17). For the dust dispersion, they employed Steinbrecher's principle (Fig. 14). A mechanical indicator for the time-pressure recording of the dust explosion was flanged to the top of the vessel.

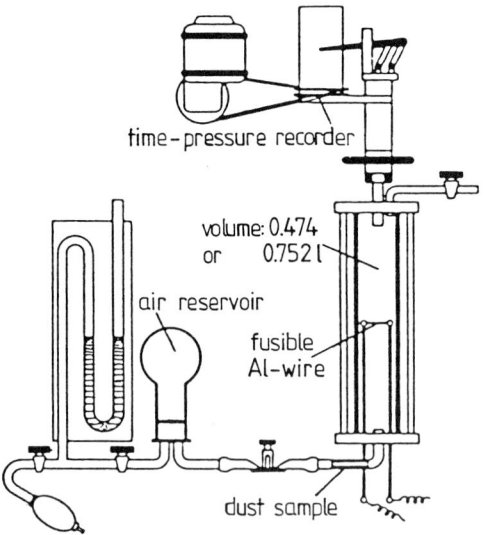

Fig. 17. Apparatus for the determination of the pressure rise of dust explosions, as per Mason and Taylor

2.3 Apparatus for the Testing of Airborne Dusts

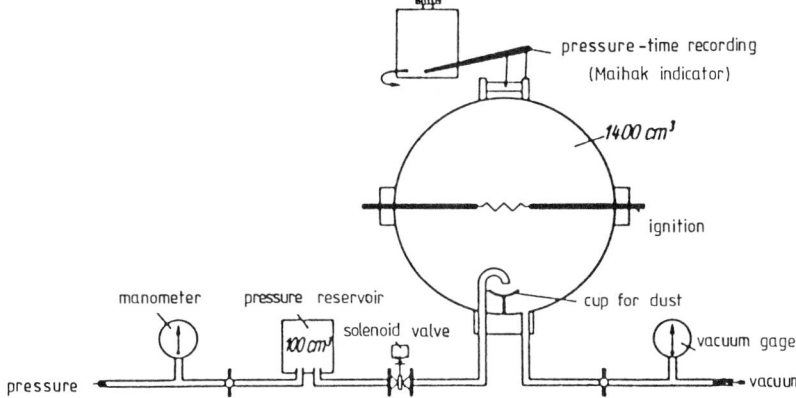

Fig. 18. Apparatus for the determination of the pressure/time behavior of dust explosions, as per Trostel and Frevert (Volume of the explosion chamber: 1.4-l)

Trostel and Frevert developed a method which directed an air jet from the top down into a cup containing the dust layer (Fig. 18). Thus, the dispersion of the dust/air mixture occurred in a spherical explosion vessel of 1.4-l volume. Again a mechanical indicator recorded the time-pressure behavior of the dust explosion.

In following Trostel's and Frevert's concept, the "Bundesanstalt für Materialprüfung" [19] (German Federal Institute for Materials Testing) used a very similar apparatus for their dust testing (Fig. 19). Initially, lead strain gages were used for the determination of the maximum explosion pressure. Once it became obvious that the flame temperature influenced the results, it was decided to use a mechanical indicator for pressure recording. As the dust dispersion method generates a slight pressure in the explosion vessel, a negative pressure was set at the start in order to arrive at 1 bar (abs) at the initiation of the dust explosion.

Subsequently, the apparatus was modified again. Figure 20 shows the arrangement developed in 1957. The sphere was made out of metal and had a volume of 1.7-l. The pressure rise of the dust explosion was transmitted through a membrane which had a mirror attached. The deflected light beam was photographically recorded.

Homemade "capignitors" served as the ignition source; with a composition similar to thermite, these ignitors required no atmospheric oxygen. This type of ignition was considered extremely powerful since it also made hard-to-ignite dusts explode. This method of ignition became the standard for all dust tests in order to maintain equal starting conditions for dust explosions to be studied.

At approximately the same time, the US Bureau of Mines developed the so-called "Hartmann apparatus". The closed explosion chamber was cylindrical and had a volume of 1.2-l. The dust to be tested was dispersed into the chamber onto a continuous electrical spark (arc) or a glowing wire coil. The values of pressure and rate of pressure rise were recorded with either a mechanical indicator or a piezoelectrical pressure transducer [36].

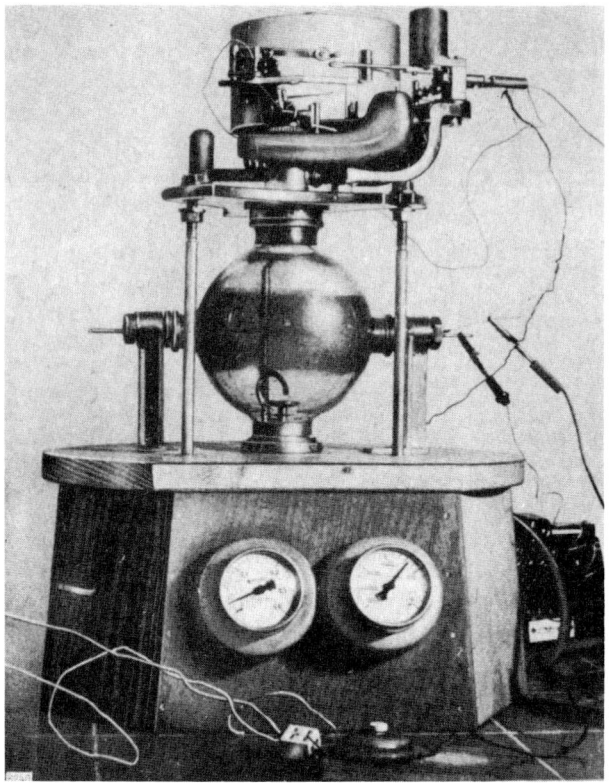

Fig. 19. Apparatus for the determination of the pressure/time behavior for dust explosions, as per BAM (Volume of the explosion chamber: 1.4-l)

Subsequently, J. Lütolf [20] simplified the Hartmann apparatus, a modification which became known as the "modified Hartmann apparatus" (Fig. 21). It is made out of pyrex glass and the violence of the explosion is expressed at two levels, depending upon the opening angle of the hinged cover. The test apparatus resembles closely the one shown in Fig. 15, which is used by the "Chemisch Technische Reichsanstalt".

However, some unavoidable shortcomings were inherent in the testing in small equipment; the safety data obtained – e.g., lower explosible limit, explosion pressure, and rate of pressure rise – did not explain the effect dust explosions had in industrial practice. Therefore, a theoretical method was developed which allowed the calculation of the maximum explosion pressure of a dust on a thermochemical basis from the combustion temperature [10]. According to this, the maximum explosion range of most dust had to be expected to be in the range of 8–12 bar gage. However, the values obtained from the small-scale tests barely reached 50% of the theoretical value.

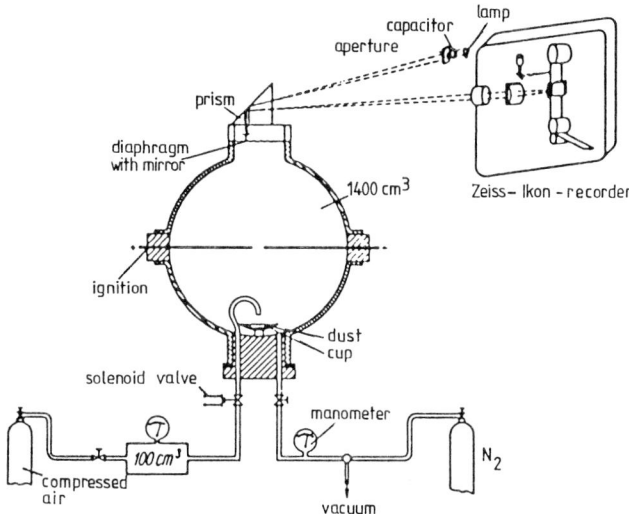

Fig. 20. Improved BAM-apparatus with a metallic sphere and optical recording unit for dust investigations (Volume of the explosion chamber: 1.7-l)

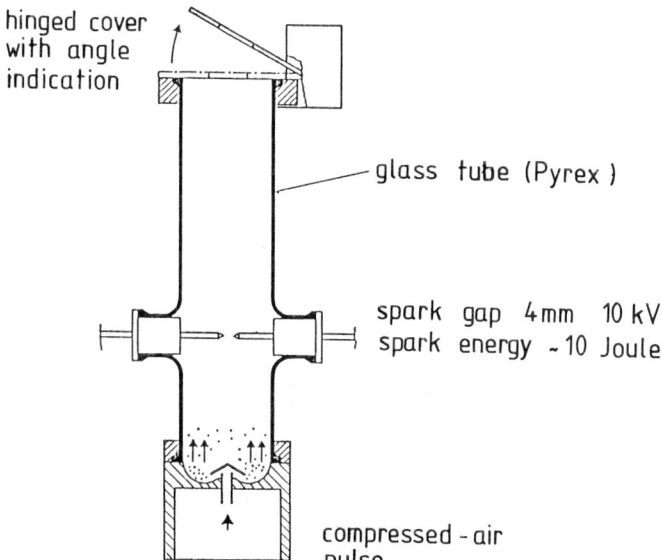

Fig. 21. "Modified Hartmann apparatus" for dust investigations, as per Lütolf (Volume of the explosion chamber: 1.2-l)

A better correlation of the values seemed to require increasing the size of the explosion vessel and improving the dust dispersion.

In 1938 Gliwitzky took the first step towards a larger test vessel [8, 9] by using a 43-l explosion vessel for his tests (Fig. 22). Impellers dispersed the dust. He was also one of the first to recognize that the activation of the ignition source had to be synchronized with the dust dispersion in order to arrive at reproducible results. The ignition source was a wire bridging two electrical terminals.

Such an apparatus was used to determine the explosion limits and the rate of pressure rise of dust explosions. The values gained from this equipment for the maximum explosion pressure for combustible dusts, e. g., aluminum, approached very closely the theoretical ones.

Fig. 22 a/b. Apparatus for the determination of the explosion limits and the rate of pressure rise of dust explosions, as per Gliwitzky; **a:** equipment/**b:** ready for testing (Volume of the explosion chamber = 43-l)

2.3 Apparatus for the Testing of Airborne Dusts

During the second World War and the time thereafter, developments ceased in the German-speaking area. It was not until 1966 that a new test procedure for dusts was developed [21], which was initially for closed vessels. It was adapted for industrial applications and gave realistic data for combustible dusts (Fig. 23). In this equipment a weighed amount of dust is charged into a dust container and maintained at 20 bar. The size of the container is in proportion to the volume of the explosion chamber. The container discharges through a fast-acting valve. Once the valve opens the dust disperses through a perforated pipe into the explosion vessel. The dust is to be ignited after a well-defined delay time t_d by the activation of an ignition source. By this arrangement, the dust is not only distributed with sufficient homogeneity, but the turbulence of the dust/air mixture at the time of ignition is almost the same for every test.

The mentioned ignition delay time t_d (the time between the start of the dust dispersion and the activation of the ignition source) coincides almost with the emptying of the dust container, so an approximation of the dust concentration is possible.

In order to reduce the risk of misjudging the explosibility of a dust, "pyrotechnic ignitors" with an energy content of 10,000 J are used as the ignition source for the tests. The ignitors weigh 1.2 g and have the following composition:

40 wt% zirconium
30 wt% barium nitrate
30 wt% barium peroxide

These ignitors ensure that hard-to-ignite dusts will be recognized as explosible.

The test method outlined above will be explained later in more detail, since it is also suitable for vessels ≥20 l and pipelines.

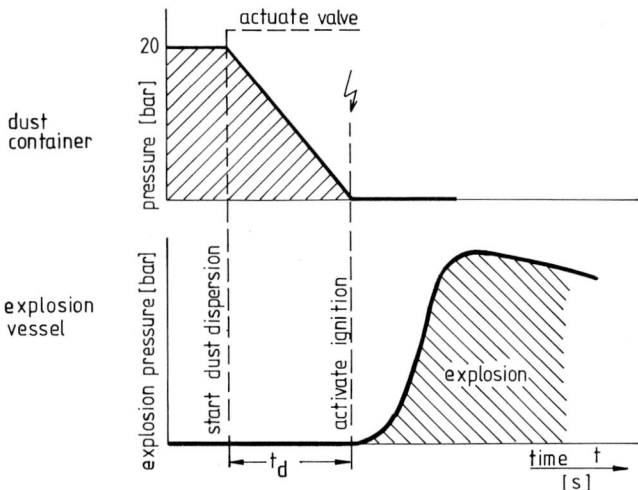

Fig. 23. Schematic representation of the dust testing procedure developed in 1966 (Volume of the explosion chamber ≥20-l)

3 Dust as a Dispersed Substance

All dispersed dusts which contain carbon can be ignited under certain conditions and are therefore explosible in terms of industrial practice [3]. Because of its reduced particle size [22], every dust is distinct from the same substance in solid or compact form.

This qualitative statement leads to a quantitative determination of the particle size, i.e., the particle size distribution of a given dust sample. In industrial applications, dust is generated as the intended end product or as an unwanted waste product.

The useable products come from particle size reduction, which can involve:

1) spraying and evaporation of liquids containing dissolved solids,
2) filtration, and
3) centrifugal separation.

Waste dusts occur in generating useable dusts because of friction and also while machining compact materials [23].

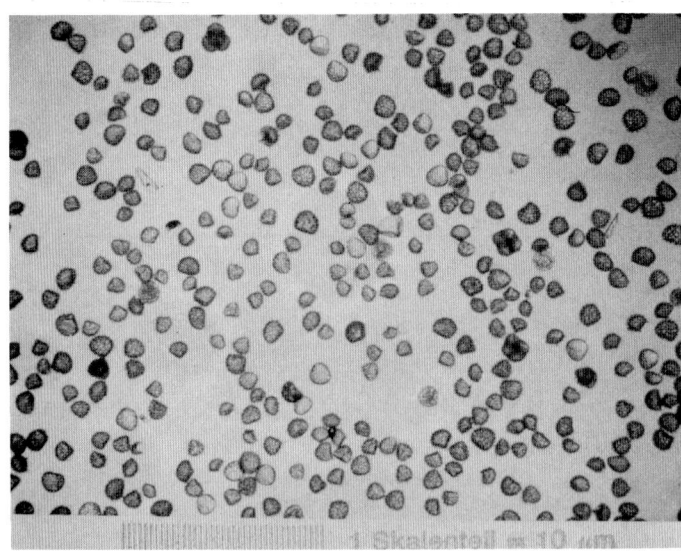

Fig. 24. Dust from Lycopodium (magnified: 1 scale = 10 μm)

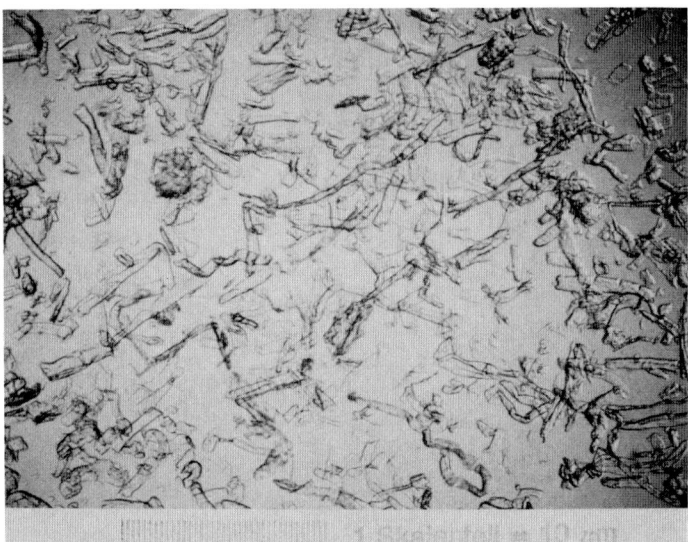

Fig. 25. Dust from cellulose (magnified: 1 scale = 10 μm)

Single dust particles can take extraordinary shapes. The external form can be granular (Fig. 24), flat and fibrous, pointed, polygonal, jagged, etc.

The size of particles varies greatly in mixtures. The largest particle can be a million times bigger than the smallest one. Therefore, the number of particles per unit volume cannot be used to characterize the dust.

A dust which is homogeneous with regard to its particle size is practically non-existent, and it is also very difficult to artificially generate such a dust.

The inherent motion of swirled-up dust creates local changes of concentration in time (separation). The solid particles settle differently depending upon their size. Gravity is the most important force for the settling of a given particle. The free-falling particle is not constantly accelerated, but because of its friction reaches a constant terminal velocity quite rapidly. This velocity can be approximated in accordance with the "Stokes' Law". Figure 26 shows the correlation of settling rate with particle size at room temperature and different densities.

The above conclusion led the committee of specialists for dust technology in the VDI to the following definition for the term dust: Dust consists of solid particles which, due to their small size, will reach a uniform settling velocity in still air after a short distance of acceleration. The velocity is approximately 0.03–100 cm/s, which is much smaller than the one expected from the "gravitation laws."

A short time thereafter [24], the committee on "Combustible Dusts" gave the following definition: "dust is a dispersed solid of any form, structure, and density."

In practice, it is common to use the term "dust" if the particle size of the solids mixture is 100–300 μm. Mixtures with particle sizes of 100–300 μm are designated as "fine" dust and the ones with particle size 30–100 μm as "finest" dust. The

26   3 Dust as a Dispersed Substance

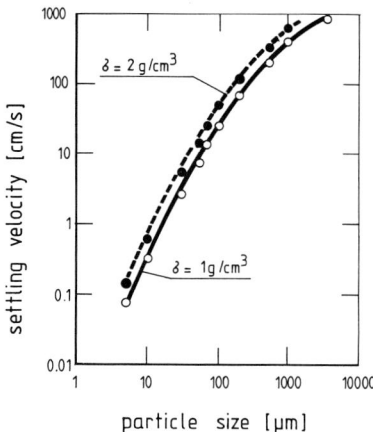

Fig. 26. Airborne dust; rate of settling as a function of particle size (room temperature)

portion of particles less than 30 μm are called by Geck [3] "dust in dust". They can remain in suspension in air for up to 48 hours.

It should be pointed out that short fibers also fall into the category "dust". Such fibers are generated by cutting synthetic continuous threads or through the grinding or processing of natural or synthetic fibrous material. Such solids (Fig. 27) are known by the term "flock", which is characterized by its dernier D equal to 1 g per 9 km of thread and the length 1 of the cut. Fibers, which once airborne are capable of burning, are mainly used in industrial coating installations (carpet industry, floor coverings, car supplies).

Fig. 27. Flock (magnified: 1 scale = 40 μm; 7.4 D; l = 0.9-mm)

4 Material Safety Specifications

4.1 Preliminary Remarks

The fire and explosion hazards which may exist while handling combustible dusts are much less known than the ones which are present with gases and solvents. Therefore, quite often a misjudgement is made. In order to safely handle combustible dusts, it is important to know the dangerous properties [23]. The most reliable way to gain information with regard to the fire and explosion behavior of a dust is to analyze a sample and to describe the test result in the form of parameters known as "safety characteristics".

Such characteristics are in general quantitative statements of product properties which should allow a judgement of the dangers inherent in chemical products or mixtures from reaction.

With few exceptions, the safety characteristics are normally not physical constants but agreed upon "conventional" values. Their significance and reproduction are tied to a specific test method, which has to be practical or allow an easy conversion of the test results into practice.

In general, sources should identify the test method which was used to determine the safety characteristics. It would be desirable to have an internationally standardized test method to enable meaningful comparisons [26].

In order to sufficiently survey the danger potential of a dust, one is normally forced to carry out a multitude of tests. Depending upon the effort spent, a more or less comprehensive picture will result, as in a mosaic. The scope of the tests has to be determined in close collaboration between tester and user, working towards the solution of the problem. It will also depend upon the safety measures selected to prevent fires or explosions or to limit the results of explosions. In the following paragraph some selected test methods will be described, along with the resulting safety characteristics, plus the influencing parameters which are significant for practical applications. A distinction will be made between "dust layers" and "airborne dust".

4.2 Material Safety Specifications of Dust Layers

(G. ZWAHLEN)

4.2.1 Flammability

Dust layers are considered to be flammable if they can be ignited with an outside source and if the fire propagates to a certain extent once the source is removed. Various ignition sources are used in preliminary tests to determine the general behavior of dust layers with respect to flammability and combustibility. The sources are: mechanically produced sparks (lighter with flintstone), glowing cigarette, burning match, and gas flame. They often have played an important role in actual mishaps (Fig. 28) [27].

A test may be repeated at an elevated temperature (e.g., at 100°C), because dusts behave differently at various temperatures (see sect. 4.2.2).

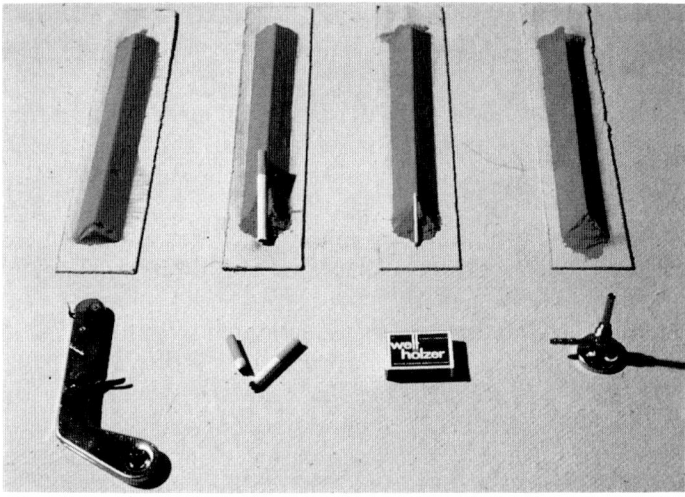

Fig. 28. Flammability of dust layers

4.2.2 Burning Behavior

Not only is a combustible dust's behavior towards various ignition sources important, but also the appearance of the fire and how fast it propagates after ignition. Simple tests give answers to these two criteria: the "combustibility test" and the test for "burning rate".

The tests are carried out in a ventilated laboratory hood. The air velocity at the test location is approximately 0.2 m/s, (in the same direction as the propagating reaction). The superimposed light air stream is necessary to vent the inert gases which are generated upon ignition of the product and which may inhibit the combustion behavior of the dust.

4.2.2.1 Combustibility Test at Room Temperature

Approximately 5ml of the ground and dried product are deposited on a heat-resistant plate (preferably ceramic) as a strip 4-cm long and 1–2-cm wide. An electrically heated, glowing platinum wire having a temperature of approx. 1000°C is dipped into the end of the deposit for approx. 5s (Fig. 29).

The behavior during combustion is rated in accordance with the course of reaction. The rating of the "class (CL)" is defined in Table 4 [28].

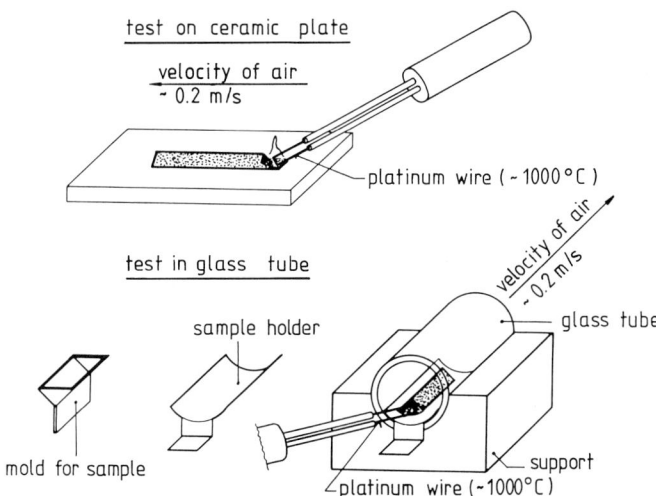

Fig. 29. Combustibility test

Table 4. Combustibility test (test results per combustion class [CL])

Test result		CL	Reference product [20°C]
No ignition		1	Table salt
Brief ignition rapid extinction	No spreading of fire	2	Tartaric acid
Localized combustion or glowing with practically no spreading		3	D + Lactose
Glowing without sparks (smoldering) or slow decomposition without flame		4	1-Amino-8-naphthol-3,6-disulfonic-acid (H-acid)
Burning like fireworks or slow quiet burning with flames	Fire spreads	5	Sulfur
Very rapid combustion with flame propagation or rapid decomposition without flame		6	Black powder

4.2.2.2 Combustibility Test at Elevated Temperature

If the combustibility of a product is of interest at elevated temperatures (e.g., if drying is contemplated), the combustibility test can be repeated at elevated temperature (e.g., at the anticipated drying temperature). This is not necessary if the test already resulted in a class 6 at room temperature. Sometimes there is a marked difference in the combustion behavior (Table 5). Therefore, the test temperature has to be listed together with the combustion class.

The testing can be done in accordance with Sect. 4.2.2.1 except that the sample with its holder (e.g., ceramic plate) is preheated in a laboratory oven to the desired test temperature (e.g., 100°C). The sample is ignited with the glowing platinum wire directly in the laboratory oven. Ample air flow has to be guaranteed.

Table 5. Examples of varying combustion classes (CL) at 20°C and 100°C

Product	CL 20°C	CL 100°C
Toluene-4-sulfonic acid chloride	2	5
Tetrahydrophthalic acid anhydride	2	5
Hydroxyquinaldine-4-carbonic acid	3	5
1-Diazo-2-naphthol-4-sulfonic acid	2	6
Dextrin	3	5

In practice, a special test arrangement has been proven successful which ensures enough air and removes the combustion gases. Additional tools are necessary, such as: a mold for the sample, sample holder, glass tube, and a heat-resistant insulated support (Fig. 29).

The sample product is loosely filled into the mold and then covered with the sample holder, so that the product rests in the center of the holder. The mold and the sample holder are turned 180° and the mold lifted off. The sample holder with the product is then inserted into the glass tube which sits on the support.

The complete assembly consisting of support, glass tube, and sample holder is preheated in the oven for one hour at the desired temperature.

Afterwards, the complete assembly is removed from the oven and ignited with the glowing platinum wire [29].

4.2.2.3 Burning Rate Test

The combustion classes in Table 4 characterize the ease of ignition of the product plus its flame appearance and propagation; but they give no indication of how fast the fire spreads after ignition. This requires a burning rate test. Using a special mold (Fig. 30), a 25-cm-long strip of the dried, ground product is deposited on a fire-resistant plate. The strip is ignited with the glowing platinum wire at one end. The time it takes to burn through 20-cm of the strip is measured with a stop watch. (The time measurement starts after the first 3-cm of the strip are consumed). This test can also be carried out at elevated temperature.

A product is classified as easily ignitable with a high burning rate if the flame takes less than 90 seconds to consume the 20-cm of the product strip [30].

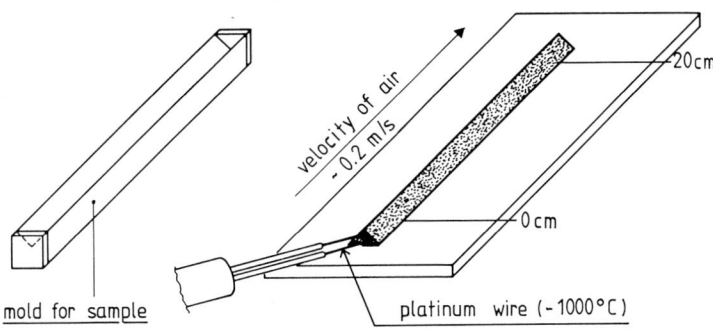

Fig. 30. Burning rate test

4.2.3 Deflagration

There are dusts which will decompose in an oxygen-deficient atmosphere, once they are ignited by an outside source. The decomposition spreads through the total amount of product more or less rapidly and produces a high temperature rise, liberating a substantial amount of gases. Such a behavior is called *deflagration*, in line with the nomenclature used in the technology of explosives.

A deflagration, for example, can be initiated by a foreign particle which has been heated up through friction in a mill and then discharged with the product into a container. It can also develop from a localized fire, which can change to a deflagration because of insufficient air infiltration.

The course of a deflagration cannot be stopped through inerting or choking.

The residue of a deflagration (ashes) can still be combustible. If the hot ashes from a deflagration are exposed to air, they may burst into flames spontaneously like an explosion (secondary mishap).

A deflagration liberates a large amount of gases, which, if not vented, will result in a pressure build-up in a closed vessel. This may cause the vessel to tear or rupture.

In the deflagration product, the transport of heat occurs primarily through convection of the hot gases from decomposition.

The spreading of the exothermic decomposition can be slowed or even prevented through an endothermic change such as melting of the product or evaporation of residual moisture. Laboratory tests have been developed in order to detect the tendency for deflagration of a product. All products should be subjected to such a test: 1 – if they have a combustion class ≥ 4 at room or elevated temperature (Table 4) or 2 – when tested for thermal stability (see Sect. 4.2.6.1) they show a spontaneous, strong exothermic decomposition.

The chemical structure of a product may also give some clues with regard to its tendency to deflagrate, e.g., if the molecule contains more than one nitro group or if it contains elements which are known to beless stable [31].

4.2.3.1 Screening Test for Deflagration

Such a test is generally used in two situations: 1 – where the sample size is limited (5–50 g) or 2 – with a product having a combustion number ≥ 4 in conjunction with the combustibility test (see Sect. 4.2.2.1). This is relevant if the reduced air pressure is of interest which will still sustain a spreading of the locally started fire. Deflagration exists if, under full vacuum, the outside ignition still results in a propagating decomposition. Such a test is also called, not quite correctly, combustibility test under vacuum. But the results are only reliable if the capability for deflagration is confirmed otherwise [32].

A dish containing 5–10 ml of the test sample is placed into a Witt'scher Pot (volume approx. 1-l). The product is ignited with a stationary glowing platinum wire. Once the sample is burning, the pot is evacuated (water ejector). If the product continues to react under vacuum, then a deflagration is at hand.

4.2.3 Deflagration

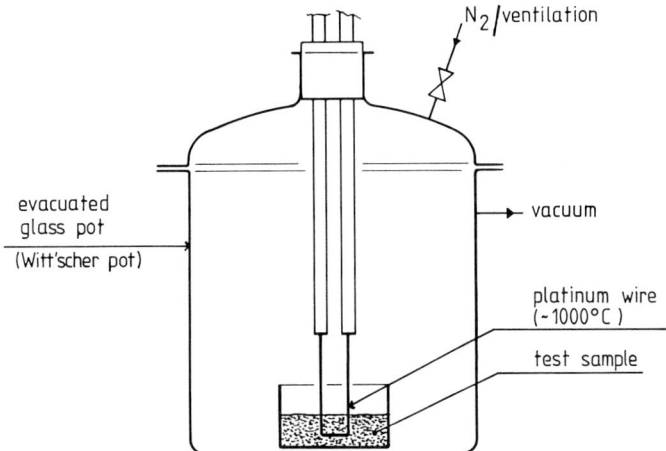

Fig. 31. Screening test for deflagration

The test is repeated with the ignition applied after the pot has been evacuated. If the product catches fire under vacuum and the decomposition propagates, then a deflagration is obvious.

4.2.3.2 Laboratory Test for Deflagration

Based on experience, in most cases, the following test arrangement gives reliable results for assessing the susceptibility of the product to deflagration. It renders quantitative results for the propagation of the decomposition front and the rise attained in temperature [33].

Approximately 200 ml of powdery substance are introduced into a glass tube and slightly compacted through tapping. After a sufficient purge with nitrogen, argon, or another inert gas, the glowing coil is switched on (for approx. 15–30 s). This triggers a local ignition. The advancing decomposition front is observed and the temperature gradient recorded at three spaced measuring points.

In case of a deflagration, the three thermocouples respond consecutively. In general, temperatures of 400 °C will be reached.

The staggered temperature curves give an indication of the deflagration velocity.

The test for deflagration is normally done at room temperature and in case of a negative result repeated at an elevated temperature (e.g., at 100 °C). For this purpose, all the test equipment with the sample is set up in an oven and preheated to the required temperature level before the test is started. Preliminary testing can be done without inerting. In such a case, a negative result is to be accepted. With a positive result, it is recommended that the test be repeated after inerting, despite the fact that the gases produced probably replace the air and create an inert atmosphere.

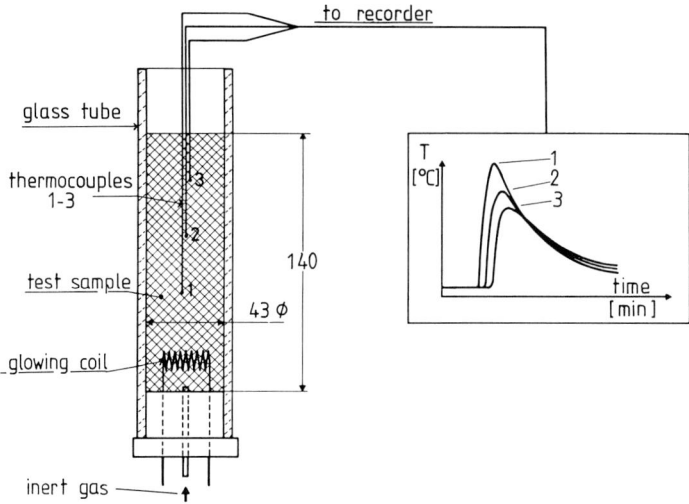

Fig. 32. Test for deflagration: spreading of a locally started decomposition

4.2.4 Smolder Temperature

The temperature of smoldering dust is the characteristic which describes the danger of ignition of a flat layer of dust on a hot surface.

It is defined [34] as the lowest temperature of a heated free-standing surface which is capable of igniting a 5-mm-thick dust layer. With thicker layers, smoldering may start even at a lower temperature. The smolder temperature can only be determined for materials which will not melt or evaporate before reaching the required temperature. The smolder temperature of a product depends not only on its chemical structure and composition but also on its physical characteristics such as particle size and bulk density. Based on experience, the smolder temperature decreases with decreasing particle size (Tables 6 and 7).

In order to properly interpret the results from the smolder temperature test, the particle size, the bulk density, and the height of the layer have to be stated.

4.2.4.1 Determination of the Smolder Temperature

The IEC (International Electrotechnical Commision) has proposed a standard for the determination of the smolder temperature. However, such a standard has not yet been adopted [36]. The test apparatus consists of a circular electrically heated metallic plate, with a diameter of 200-mm and a thickness of >20-mm, with temperature control and recording equipment for the plate and sample temperatures. The plate must be capable of reaching 400°C without a dust layer.

4.2.4 Smolder Temperature

Table 6. Various products with smolder temperatures (taken from VDE 0165/8.69)

Product	Max. particle size [μm]	Maj. portion having particle size [μm]	Bulk density [kg/l]	Smolder temperature [°C]
Phosphorus	150	30– 50	0.99	305
Iron powder	500	100–150	1.6	240
Rye flour	200	30– 50	0.31	325
Wood flour (pine)	150	70–100	0.22	315
Charcoal	20	1– 2	0.36	340
Naphthalene	300	80–100	0.53	Melted
Polyvinyl chloride	10	4– 5	0.55	Charred

Table 7. Correlation of smolder temperature with particle size and height of dust layer [35]

Dust	Sieve fraction	Smolder temperature °C for a	
		3-mm high layer	6-mm high layer
Flame coal (bituminous coal)	<200 μm < 70 μm	295 270	270 230
Gas coal (bituminous coal)	<200 μm < 70 μm	400 270	270 245
Lean coal (semianthracite) (subbituminous)	<200 μm < 70 μm	450 340	350 280

With the help of a ring having an ID of 100-mm, a 5-mm-thick layer of dust of known particle size and density is formed on the hot plate. The sample temperature is measured with a thermocouple which is strung parallel to and 2–3-mm above the hot plate surface through the dust layer. The measuring point is in the center of the plate (Fig. 33). A temperature measurement without contact is also possible.

The temperature is maintained constant for the hot plate. The lowest temperature value is recorded which causes the deposited sample to smolder, glow, or burn. A temperature rise of >20°C, measured by the thermocouple buried in the dust layer, is interpreted as equivalent to an igniton.

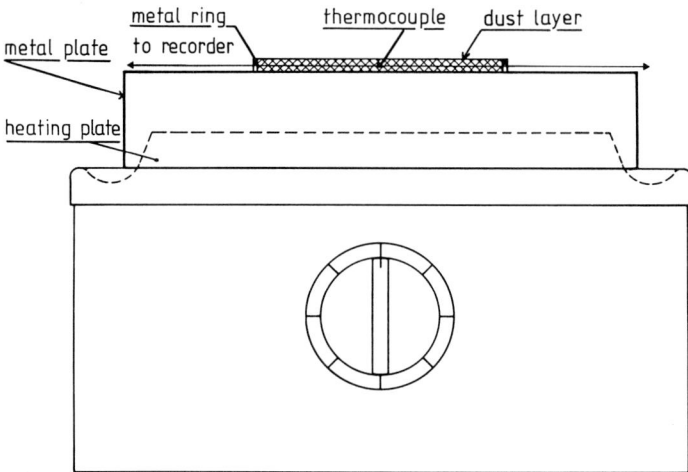

Fig. 33. Apparatus for the determination of the smolder temperature (schematic)

4.2.5 Autoignition

It is understood that autoignition means the ignition of combustible matter in air subjected to uniform heat. The temperature of the surrounding atmosphere (storage temperature) which initiates autoignition after self-heating of the product is called the autoignition temperature.

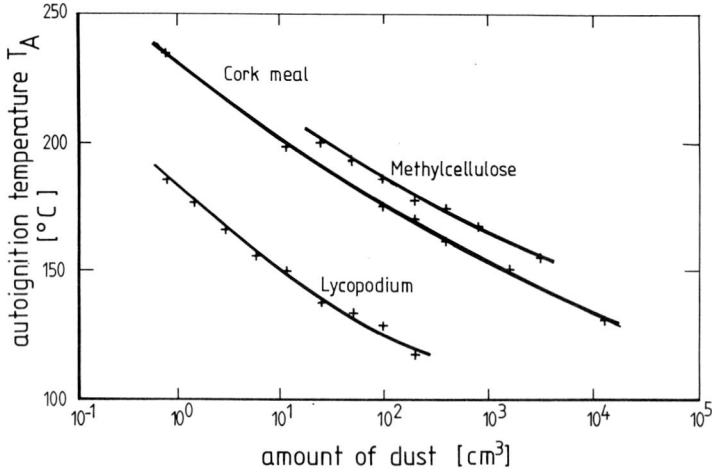

Fig. 34. Correlation of autoignition temperature of cylindrical dust piles with dust quantities [27]

The self-heating of the product is caused by an already evident oxidation in air at storage temperature. Such oxidation will liberate a certain amount of heat per time and mass. If the heat is not entirely transmitted to the surroundings, an increase in the temperature of the product will occur. The end result is autoignition.

Whether there will be self-heating or even autoignition depends upon the ratio of heat generation to heat transmission. The ambient temperature which will initiate the autoignition of a certain dust will therefore not only depend upon the type of dust (chemical nature, particle size, moisture content, ...) but also upon the storage mode, the shape and size of the pile, and the duration of heat exposure.

All statements for the autoignition temperature have to include the shape and size of the dust pile (method dependency).

4.2.5.1 Determination of the Relative Autoignition Temperature, as per Grewer

The apparatus has an air-purged oven with 6 wire mesh baskets (Figs. 35 and 36) plus temperature control and recording of the oven and probe temperature. Five different substances can be tested simultaneously [37].

a) Determination under temperature-programmed conditions
Approximately 7–8 ml of the ground and dried test substance and the same amount of graphite dust are heated 1°C/min in a hot air stream (2.0 l/min) up to 350°C. The samples are kept in a cylindrical basket made of a fine wire mesh. The temperatures of the samples are recorded.

A crossing of the temperature curve of the sample with the one of the reference substance is interpreted as an exothermic reaction.

The temperature level at the crossing of the sample curve with the reference curve is stated.

If the sample temperature reaches a level of $\geqslant 400$°C, this is classified as autoignition; otherwise it is called self-heating [38].

b) The same apparatus can be used for the determination of the autoignition temperature under isothermal conditions, i.e., at constant oven temperature.

The lowest oven temperature is found which will result in an exothermic reaction. The isothermal test renders, in general, a lower temperature than the test with the programmed temperature rise. Therefore, the test reports have to include the test method and procedure.

c) The official publication of the European Community describes a modified test method for the determination of the relative autoignition temperature [38]. Instead of a cylindrical basket, a cubical basket is used as the sample container. The oven is also of a slightly different construction. Up to now, the results obtained with the EEC method have been equivalent to those of Grewer.

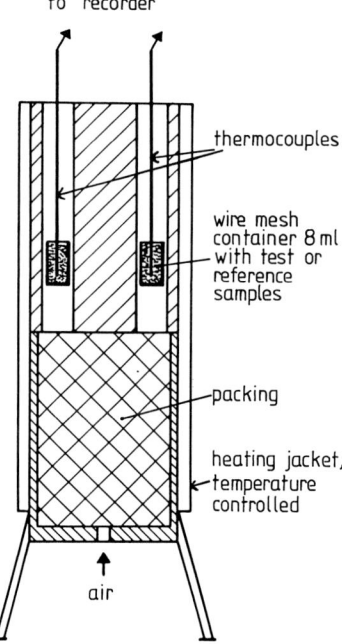

Fig. 35. Apparatus for the determination of the relative autoignition temperature, as per Grewer (schematic)

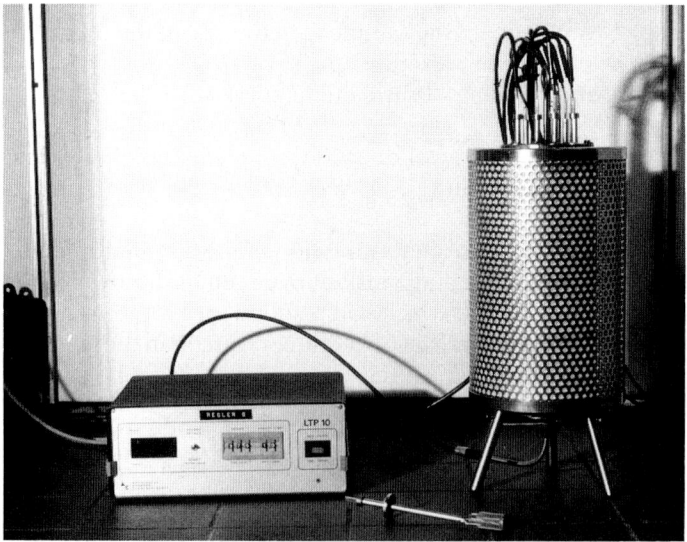

Fig. 36. Apparatus for the determination of the relative autoignition temperature, as per Grewer (actual)

4.2.5.2 Hot Storage Test in the Wire Mesh Basket

A so-called hot storage test can be used to assess the dangers which may exist in handling larger product quantities [27]. The products to be tested are filled into cylindrical wire mesh baskets (diameter = height) which are stored in an air-purged oven (2 l/min) at constant temperature (isothermal conditions). Normally, a 400 ml wire mesh basket is used; on occasions, 1600 ml or larger units are used. In general, the relative autoignition temperature decreases with increasing volume (Fig. 34). Again the temperature curve of the test product is compared with the reference material (graphite).

The test is repeated with fresh samples at different temperature levels until there is no exothermic reaction for 24 hours. In case of an exothermic reaction at a given oven temperature, the duration of the test is extended until the maximum of the self-heating is reached or passed.

The report must include: sample volume, oven temperature, maximum sample temperature, induction time (the time it takes for the sample to attain the peak temperature after it has reached oven temperature). In case there is no self-heating, the total duration of the test is recorded. Depending upon the application, it may be beneficial to plot the results. Such plots include the correlation of autoignition temperature with sample size (at constant storage temperature) (Fig. 34), a plot of induction time versus storage temperature (at constant volume), or induction time versus sample volume (at constant storage temperature).

Such plots facilitate the scaling of test results for actual process conditions.

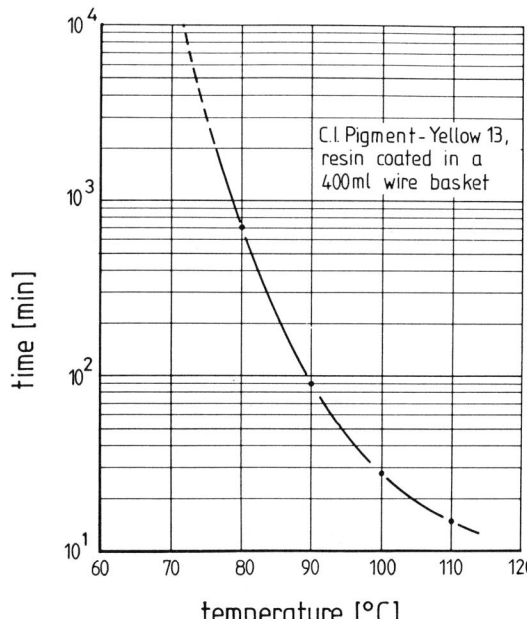

Fig. 37. Induction time versus storage temperature (at constant volume)

4.2.6 Exothermic Decomposition

Elevated temperatures may subject a product to a chemical transformation which, contrary to self-heating (see Sect. 4.2.5), will not require oxygen. The reaction may be endothermic or exothermic. An exothermic reaction is classified as an exothermic decomposition. It is, from a safety point of view, of great importance, contrary to endothermic decompositions.

Such a reaction may liberate gases of decomposition (smolder gases), which will result in a pressure rise in a closed vessel, which may subsequently tear or burst.

In addition, the gases of decomposition may be flammable and present an explosion risk.

In case of an exothermic change, the heat may be trapped, resulting in self-heating, with a thermal explosion as the consequence. As with autoignition, the exothermic decomposition depends upon the volume and size of the sample. With increasing volume, the danger of a heat accumulation increases.

The decomposition temperature measured is a relative value. It is dependent on the test method, and a conversion to actual conditions is problematic. In any case, the test method has to be stated together with the temperatures.

4.2.6.1 Determination of the Exothermic Decomposition Temperature in an Open Vessel, as per Lütolf

This determination involves the lowest temperature which results in an exothermic reaction of a product in a test tube at oxygen-lean conditions [39].

Apparatus: electrically heated oven for 6 test tubes (Figs. 38 and 39) with temperature control and recording for the oven and sample temperatures.

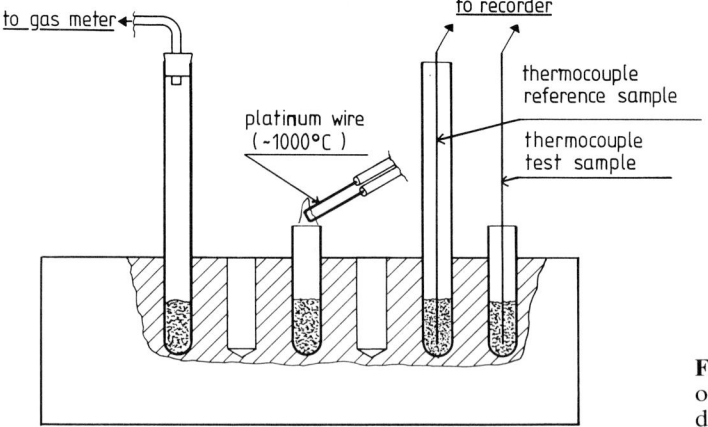

Fig. 38. Determination of an exothermic decomposition, as per Lütolf (schematic)

4.2.6 Exothermic Decomposition

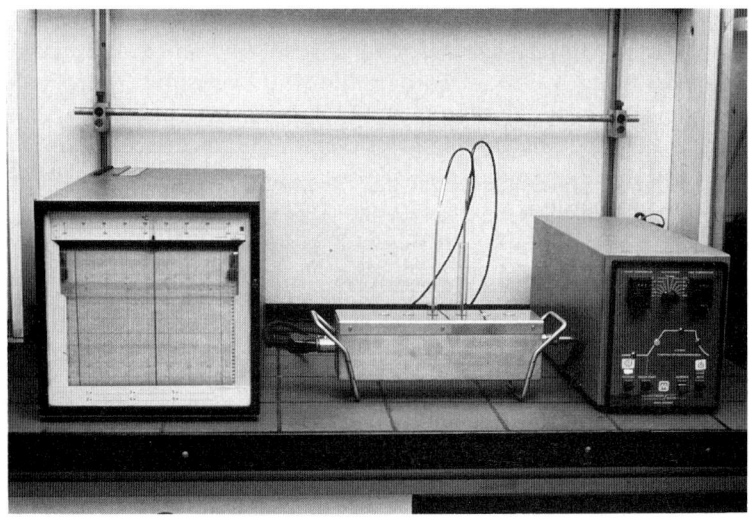

Fig. 39. Determination of an exothermic decomposition, as per Lütolf (actual)

a) Determination under temperature-programmed conditions
Approximately 2 grams of the test sample and 2 grams of graphite as a reference substance are heated in a test tube on a common heater at 2.5°C/min up to 350°C oven temperature. The sample temperature is recorded. An exothermic reaction is indicated when the temperature curve of the product is above the one for the reference substance (Fig. 40).

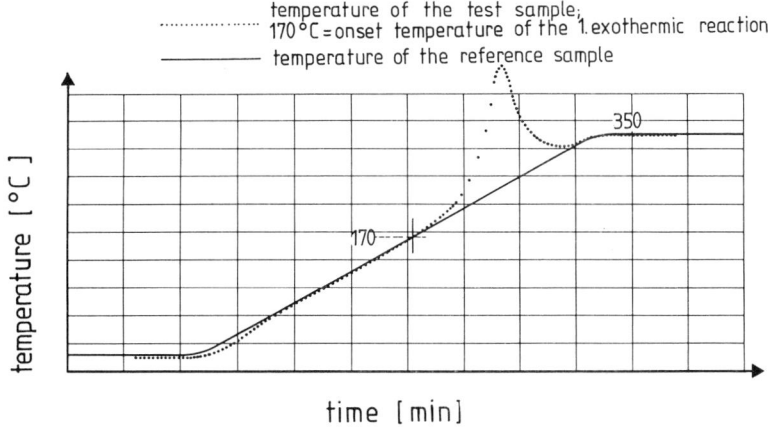

Fig. 40. Testing for exothermic decomposition under temperature-programmed conditions; course of temperature

b) Isothermal test

If the temperature-programmed test results in an exothermic reaction, then the sample is subjected to an isothermal test (at constant oven temperature). Again 2 g product are used, and the starting temperature will be the one where the product recording in the programmed test crosses the reference curve. The isothermal test is repeated with fresh samples in decreasing 10 °C steps until the substance does not show an exothermic reaction for 8 hrs (in certain cases even longer).

With such a test, two distinct behaviors are typical:

1) The violence of the exothermic reaction decreases with decreasing oven temperature for the majority of the tested products (Fig. 41).
2) A delayed exothermic reaction may be noticed with a minority of products. With decreasing oven temperatures the time delay for the start of the exothermic reaction will increase, but the magnitude and the violence will hardly be affected over a wide temperature range. This is called autocatalytic behavior (Fig. 42).

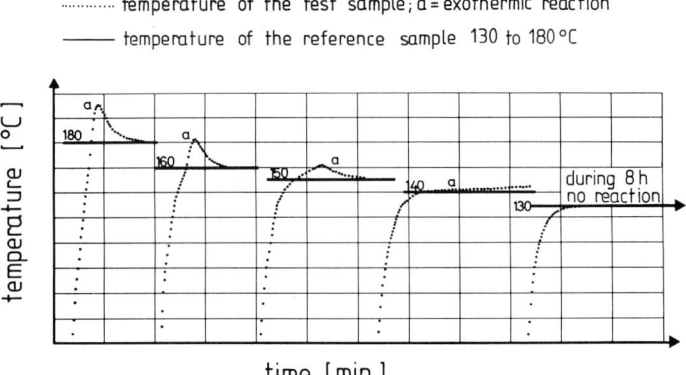

Fig. 41. Isothermal test for exothermic decomposition. Normal temperature behavior

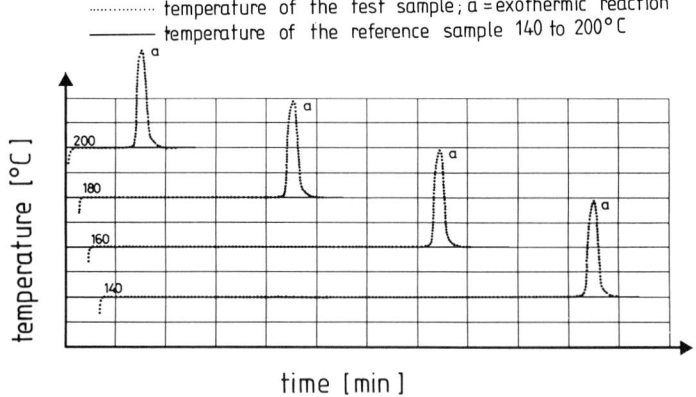

Fig. 42. Isothermal test for exothermic decomposition. Autocatalytic temperature behavior

4.2.6 Exothermic Decomposition

Products which show an autocatalytic behavior have to be analyzed very carefully from a safety point of view since the exothermic decomposition may start after a time delay of days or weeks under actual conditions.

c) Testing the influence of materials of construction on the exothermic decomposition

Proceed as per Sect. 4.2.6.1 a) and b). 100 mg of the material of construction are added to the sample. For instance, iron powder or stainless steel turnings. Compare the results obtained with and without the additional material.

d) Testing of the flammability of the gases of decomposition

Proceed as per Sect. 4.2.6.1 a). While heating up, check the flammability of the gases or vapors produced at a certain temperature interval (e.g., 50°C) with an ignition source (glowing platinum wire, gas flame, match) (Fig. 38).

e) Determination of the amount of decomposition gases

Proceed as per Sect. 4.2.6.1 a) with the following variation: A test tube containing 1 g product is sealed with a rubber stopper equipped with a glass pipe with latex tube. The test tube is put into a heating block at 350°C.

The gases are passed through an empty wash bottle to a gas meter (Figs. 38 and 43). In general, gas production stops after a few minutes. The gas quantity is stated as l/kg.

In special cases, the gas evolution is also measured under different parameters such as isothermally at 250°C over 8 hrs.

The results obtained from the outlined test methods, which are relatively insensitive, can only be transposed to actual conditions with an adequate safety margin.

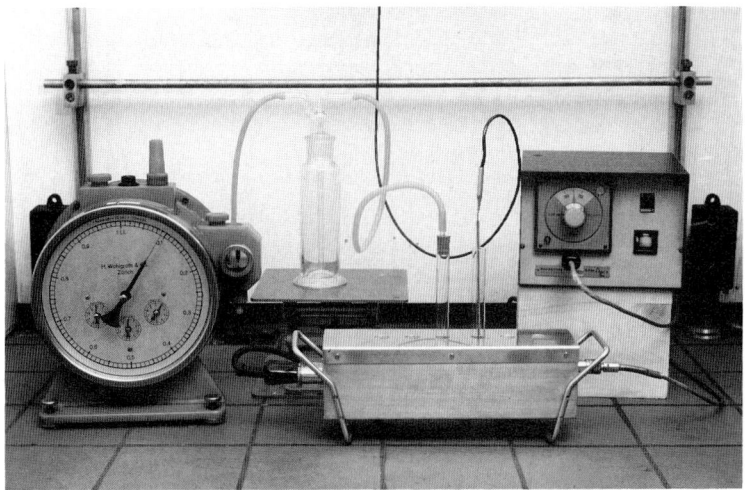

Fig. 43. Exothermic decomposition, measuring the gas evolution

4.2.6.2 Determination of an Exothermic Decomposition in an Oven Purged with Nitrogen, as per Grewer

The approach and mode of operation are the same as outlined in Sect. 4.2.5.1: determination of the relative autoignition temperature as per Grewer. But instead of air, a nitrogen stream is passed through the oven.

4.2.6.3 Differential Thermal Analysis

This is a sensitive microthermal analytical method (micro-calorimetry), which determines quantitatively the heat generated in a decomposition reaction relative to time and temperature under temperature-programmed or isothermal conditions. The test is conducted in open or closed vessels with 5–50 mg product.

The quantitative micro-thermal analytical results are suitable as the base for the calculation and conversion to actual conditions. These include: heat production rate, as a function of temperature, maximum decomposition energy, time to reach maximum heat generation, etc.

4.2.6.4 Determination of an Exothermic Decomposition Under Choked Heat Flow

The testing for thermal stability is carried out under almost adiabatic conditions (the energy generated by the exothermic decomposition is not removed). Therefore, self-heating of the substance occurs, which accelerates the decomposition. The decomposition reaction is excited to the maximum heat generation. In case of a sufficiently high temperature increase, which is independent of the total energy of decomposition, there may be a thermal explosion (Fig. 44) [40, 41].

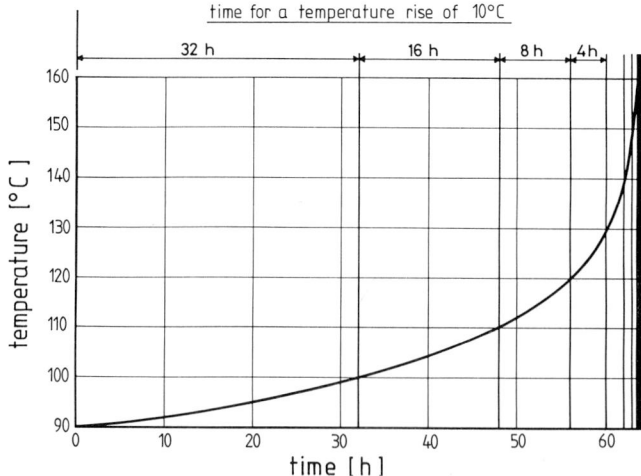

Fig. 44. Example of self-heating due to exothermic decomposition up to thermal explosion; course of temperature under adiabatic conditions

4.2.6 Exothermic Decomposition

Whether a dangerous self-heating occurs under actual conditions depends upon the ratio of heat generation to heat transmission.

As long as the capacity for heat removal is greater than that for heat generation at working conditions (storage condition) there will be no danger. Once an equilibrium is reached, the temperature will drop after a certain time.

If the heat generation exceeds the heat removal at the critical temperature of the system, there will be an accelerated decomposition reaction and temperature increase as shown in the example in Fig. 45 [42].

The generation of a large amount of flammable decomposition gases will increase the danger. It creates an additional explosion hazard which by itself may lead to a tearing or bursting of the vessel.

The course of the reaction and the critical temperature depend upon the volume of the dust pile and the heat generation of the decomposition reaction. With increasing volume, both the critical heat generation and critical temperature decrease.

Once the total energy of the thermal decomposition is large enough, it will lead to a thermal explosion, as shown for a specific example in Table 8 [42].

The test "heat accumulation in the Dewar flask" [43] is suitable for studying the behavior of a thermal decomposition under almost adiabatic conditions.

The test substance is stored in a Dewar flask in a heat box at constant temperature. Normally, Dewar flasks with a nominal volume of 200 ml are used. 500 ml, 1000 ml, and 1500 ml flasks are also possible. The sensitivity of the test increases with increasing volume.

The substance which is to be tested is filled into the flask leaving a 2-cm rim. The flask is closed with a cork which holds a thermocouple that is embedded in the

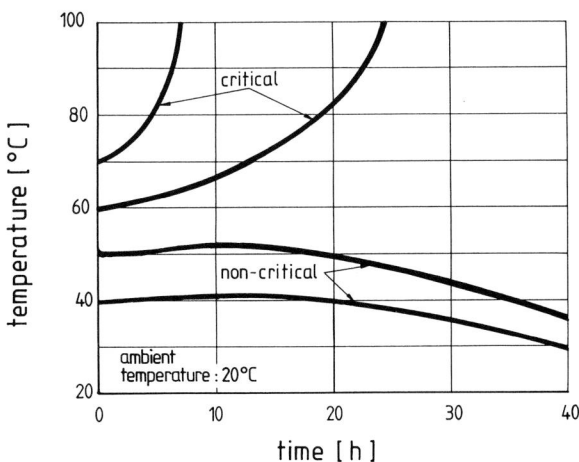

Fig. 45. Example of self-heating due to an exothermic decomposition. Temperature gradient in the center of a free-standing 150-l drum

Table 8. Example of self-heating due to an exothermic decomposition. Dependence of the critical parameters upon the volume of the dust pile

Volume [l]	Critical heat production [mW/kg]	Critical temperature of the sample reaction [°C]	Time characteristic for almost adiabatic behavior of the system [time to thermal explosion]
10000	2	37	20 days
1000	10	48	4 days
100	60	60	20 hours
10	280	73	5 hours
1	1400	86	1 hour

product. The cork stopper may be protected with a thin teflon foil. It is advantageous to preheat the product to 10–20 °C below the test temperature. The flask is put into an oven (hot box) which is preheated to the test temperature and which has vent openings. At the same time, a flask (e.g., Erlenmeyer flask) filled with graphite is put into the oven. The temperatures sensed by the thermocouples in the product and the graphite are recorded. The temperature of the graphite is the reference temperature. The self-heating of the product is indicated by a rise of the product temperature over the reference temperature. The test is repeated with a fresh sample at a 10 °C lower temperature until exothermic reaction has ceased for three days. In a special test procedure the oven temperature is adjusted to the product temperature in order to prevent heat losses. Products which develop a significant vapor pressure at the test temperature are better tested in a pressure vessel [44, 26].

The report must include statements with regard to: sample volume, oven temperature, maximum product temperature reached, and induction time (the time it takes for the sample to attain the peak temperature after it has reached oven temperature). In case there is no self-heating, the total elapsed test time is recorded.

The graph showing the correlation of induction time t_i with oven temperature t_o at constant volume expressed as $\log t_i = f(1/t_o)$ is in most cases a straight line (Fig. 46).

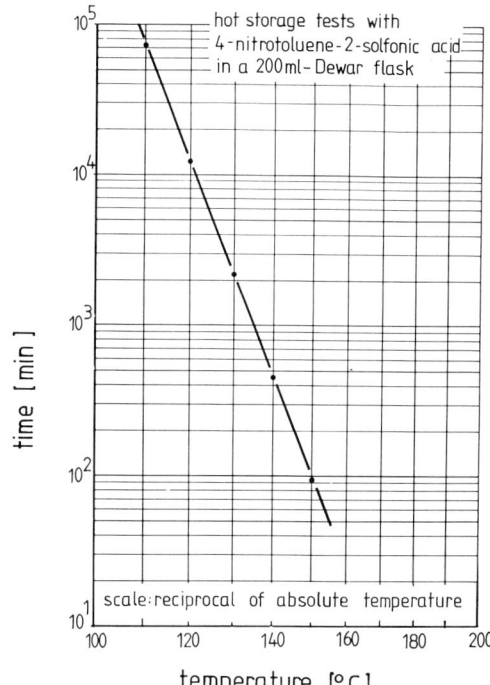

Fig. 46. Self-heating due to an exothermic reaction. Correlation of induction time with oven temperature

4.2.7 Explosibility

An explosible substance can spontaneously decompose due to temperature change, mechanical loading, shock waves from a detonation, or other reasons. Large volumes of gases will be liberated in an short time resulting in a rapid pressure build-up (explosion, detonation).

The explosion hazard of solids is judged in accordance with the results gained from the tests promoted by the European Community [45]:

a) Testing of the mechanical sensitivity towards impact loading (impact sensitivity)
b) Testing of the mechanical sensitivity against friction (friction sensitivity)
c) Testing of thermal sensitivity towards an external heat source (test with steel tube)

There is no immediate connection between the three tests. A product which is thermally sensitive, need not be mechanically sensitive and vice versa. However, most products which are sensitive to friction are also impact-sensitive. Yet there are a great number of products which are impact-sensitive but insensitive to friction.

Chances are small that a mechanically sensitive, explosible product would not be detected if only the test for mechanical sensitivity were performed with the drop weight (impact sensitivity) and no friction test (friction sensitivity) were performed. Various countries require additional testing in line with their laws covering the handling of explosives. A tabulation of the usual methods is given in [46].

4.2.7.1 Impact Sensitivity

A product is impact-sensitive and therefore explosible if it disintegrates with a bang upon its exposure to impact energy under given test conditions.

The test apparatus uses a drop weight. There are two models available:

1) weight 49 N; drop height 80-cm (as per Lütolf) [39] or
2) weight 98 N; drop height 40-cm (as per Koenen) [46, 47].

In both cases, the work done by the drop weight is 39.2 Nm (Fig. 47).

A 100-mg sample of dried and carefully ground product is subjected to the impact of the drop weight. The test is repeated ten times or until a detonation occurs (bang). With a detonation, the substance is considered impact-sensitive (Fig. 48).

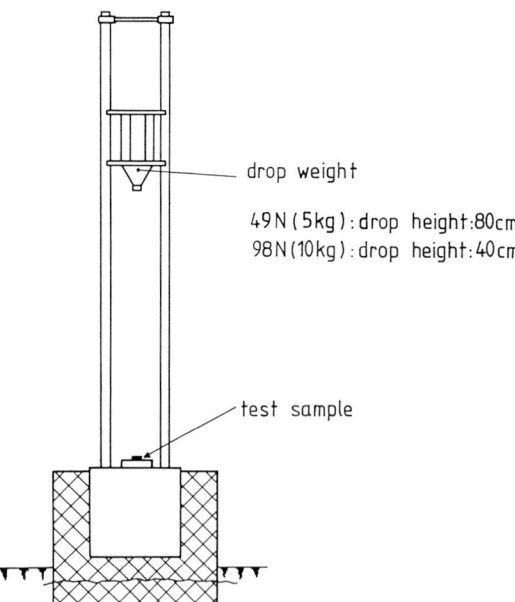

Fig. 47. Testing for impact sensitivity; drop weight (schematic)

4.2.7 Explosibility 49

Fig. 48. Testing for impact sensitivity; detonation under the drop weight

4.2.7.2 Friction Sensitivity

The test for friction sensitivity is carried out with a special device developed by the German Federal Institute for Testing Materials (BAM) [48]. The device consists of a finely grooved porcelain plate (25×25×5-mm) with a weight-loaded porcelain pin (10-mm diameter) (Fig. 49).

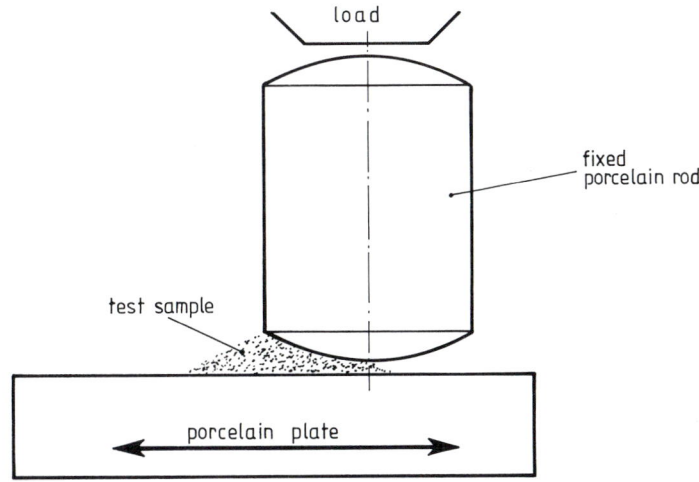

Fig. 49. Testing for friction sensitivity (schematic)

The dried product is put in between the porcelain plate and the porcelain pin and exposed to friction by moving the plate at defined conditions. The maximum speed of the plate is 7 cm/s. The sample amounts to approximately 10-mm^3. The behavior of the product under friction exposure is observed and recorded.

The intensity of friction is changed by varying the weight load on the pin. The criteria for judgment are: no reaction, partial reaction, ignition, crackling, or banging. If ignition, crackling, or banging is noticed at a load of ⩽360 N on the pin, then the product is classified as friction-sensitive and explosible.

4.2.7.3 Thermal Sensitivity

The "steel tube test" to determine the thermal sensitivity was also developed by the German Federal Institute for Testing Materials (BAM) [49].

The sample is loaded into a drawn steel tube of 30 ml content under defined conditions. The steel tube is closed through a plate with a nozzle (small hole) held by a nut.

The steel tube, which is specially supported in an protective steel chamber, is heated with four burners.

The product decomposes under the influence of the extreme heat. The liberated gases of decomposition escape through the hole in the nozzle plate. A pressure builds up if the gases cannot escape continuously, resulting in a deformation, tearing, or explosion of the steel tube. Whenever the steel tube disintegrates into at least three parts, an explosion has taken place, as per definition of the regulatory agency for explosives.

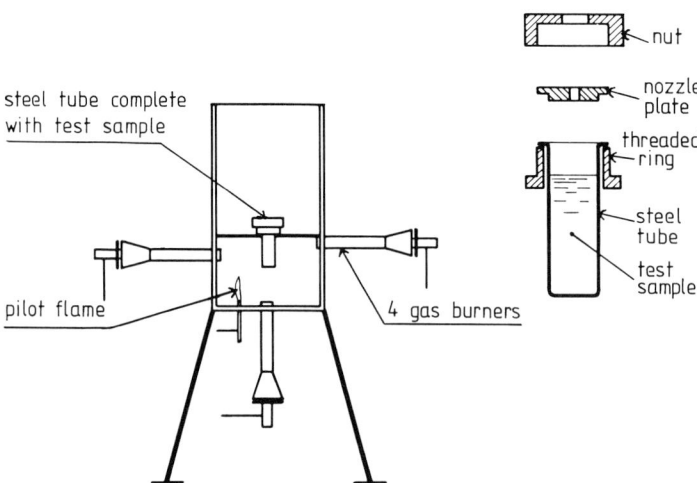

Fig. 50. Testing for thermal sensitivity; steel tube test (schematic)

The test is carried out with nozzle plates having different hole sizes.

The largest diameter of the hole is stated as the limiting diameter which causes at least three fractures of the tube in one of three tests. With a limiting diameter of ≥2.0-mm, the product is thermally sensitive and considered explosible. For transportation purposes, it is classified as an explosive, the same as impact- and friction-sensitive products are.

In order to carry out such tests, a bunker or suitable open-air spaces are needed.

4.3 Material Safety Specifications for Dust Clouds Describing the Explosion Behavior

4.3.1 Combustible Dusts

4.3.1.1 Preliminary Remarks

Dust explosions have a certain similarity to gas explosions with regard to the chemical process and results. Many phenomena which occur with gas explosions can be transferred to dust explosions. But from a physical point of view, there exists a major difference between the course of dust and gas explosions, and this creates difficulties in the research of the explosion behavior of dust/air mixtures (see sect. 2.3.).

Gas/air mixtures are dispersed in molecular form, i.e., small particles – the molecules – are next to each other and do not deviate much in size. In a dust/air mixture, however, coarse dust particles are dispersed next to gas molecules, as discussed before. The weight of the dust particle exceeds that of the gas molecule by many orders of magnitude.

The coarsely dispersed state of a dust/air mixture is not constant, as already outlined; at the very moment the mixture is produced or created, a separation starts, which is complete once everything is settled on the floor.

These are the reasons for the main difficulties encountered in experimentally investigating explosible dust/air mixtures. The means must be available to generate an almost uniform, reproducible mixture in the test vessel and to ignite the dust/air mixture always at the same level of turbulence. It is known from the studies of combustible gases (Fig. 51) that the turbulence influences not so much the pressure level but the pressure rise, which will continually increase with increasing turbulence.

This results in the constraint (Gliwitzky [9] already noticed) of synchronizing the dust dispersion with the activation of the ignition source, in order to obtain reproducible results for a given dust or comparable values for various dust types.

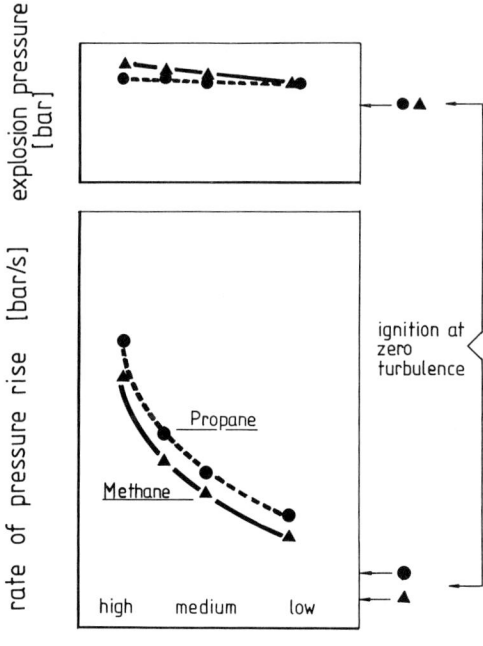

Fig. 51. Flammable gases: Influence of turbulence upon explosion behavior

4.3.1.2 Particle Size Distribution

It has already been mentioned that the particle surface area (Fig. 10) and therefore the particle size distribution within a dust type (see Sect. 3) has a considerable influence upon the combustion of a dust/air mixture. Figure 52 shows such a particle size distribution for two dusts. Such a distribution is obtained through sieving and may vary greatly.

For practical purposes, it has been expedient to specify the so-called median value M of the particle size in comparing the safety characteristics of airborne dusts. M is the 50 wt% point of the particle size distribution. In Fig. 52, Lycopodium dust has a median value of M = 32 μm, the cellulose dust M = 22 μm. In both cases we are dealing with fine dusts.

It is also clear that the median value is only descriptive and not definite. Dusts with the same origin and median value may not have the same combustion behavior, the reason being that the proportion of fine and coarse dust varies and the particle size distribution is not the same. However, from testing experience, we know that this variability becomes negligible if the median value is smaller than 63 μm.

4.3.1 Combustible Dusts

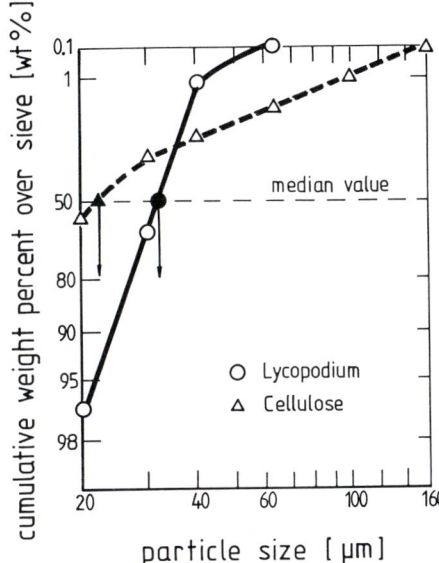

Fig. 52. Graphical presentation of the particle size distribution for two dusts

4.3.1.3 Explosibility

A dust is considered explosible if, after igniting the dust/air mixture with a suitable source, there is a flame propagation (Fig. 53) in combination with a rise in pressure.

Fig. 53. Explosion of a pharmaceutical dust/air mixture in air (4 kg product)

54   4 Material Safety Specifications

Such an explosion process occurs, as is generally known, due to mutual interaction between reactants. The products of reaction may be formed directly or through complicated intermediate steps. The required oxygen for this reaction is mostly supplied from the combustion air, with the nitrogen acting as ballast [50, 51]. It is important that the overall reaction is exothermic. For the following discussion, only dusts are called combustible in the airborne state if they require oxygen from the air for an exothermic reaction.

There exists no uniform definition for the term explosion. Only the most important will now be described.

Beyersdorfer [4] suggests: "In a narrow sense, every sudden volume increase due to a physical or chemical process is an explosion".

The VDI professional committee on "Combustible Dusts" [24] defines an explosion as "a combustion which occurs so quickly that the heat-related expansion of the participating gases results in a marked and measurable rise in pressure."

As per DIN standard 20163 [52], an explosion is "a rapid conversion of potential energy into work of expansion or work of compression, or a combination of both kinds of work with the generation of shock waves (compression shocks)."

The revised VDI guideline 2263 [23] finally states: "an explosion is a rapid combustion, with a marked and measurable pressure increase".

4.3.1.4 Explosible Limits

As already mentioned, every combustion process is an exothermic reaction. This means that a reaction initiated by a suitable ignition source produces more heat per unit of time than it consumes. The velocity at which the reaction – the flame front of the explosion – propagates through the mixture depends upon the concentration ratio of the combustible dust to air. At a certain ratio of dust and air, the normal velocity of combustion reaches a maximum; it decreases at higher as well as at lower concentrations of the combustible dust. A "lower explosible limit" (LEL) as well as an "upper explosible limit" (UEL) also exist. They are also called *flammability* or *ignition limits*. Beyond the explosible limits, the explosion reaction cannot sustain itself.

For industrial operations, the LEL is of particular interest and for a large number of industrial dusts lies between 15 and 60 g/m^3. It is not influenced by the oxygen/nitrogen ratio of the combustion air provided it is not below the so-called "limiting oxygen concentration" (LOC) (Fig. 54).

The UEL is generally very high at 2 to 6 kg/m^3. It decreases markedly at lower oxygen content of air (Fig. 54).

Basically, the lower explosible limit of a dust/air mixture can be determined experimentally (see Sect. 4.3.1.5) or through thermochemical calculation [53, 54].

Experiments have shown (Fig. 55) that the lower explosible limit of combustible dusts is independent of the "ignition energy" (IE) over a wide range (dusts A to C). Only when IE is too low to allow an explosion, will LEL increase slightly. Only in the case of dusts with relatively high LEL (dust D in Fig. 54) will IE have a strong influence on LEL.

4.3.1 Combustible Dusts

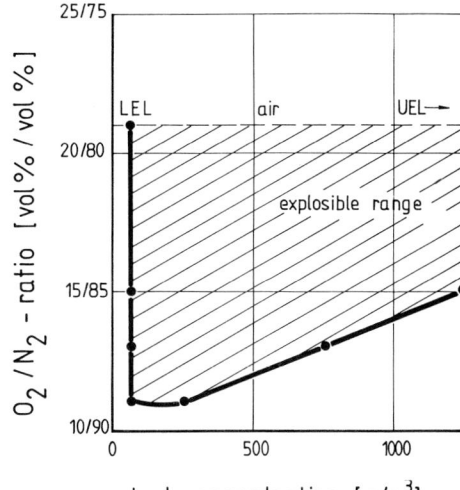

Fig. 54. Influence of the oxygen/nitrogen ratio upon the explosible limits of a combustible dust

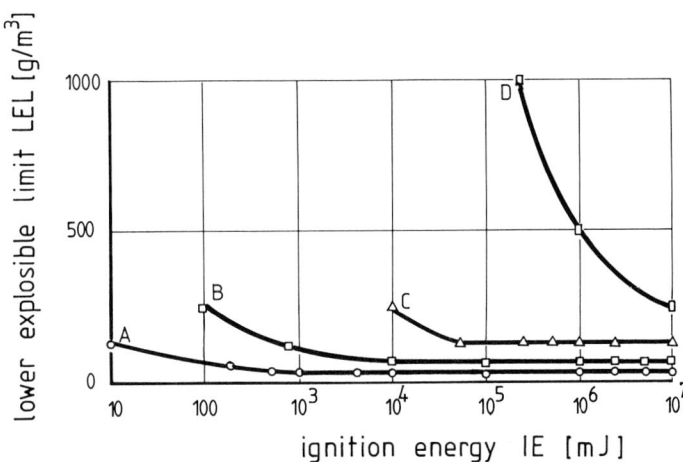

Fig. 55. Influence of the ignition energy (IE) upon the lower explosible limit (LEL)

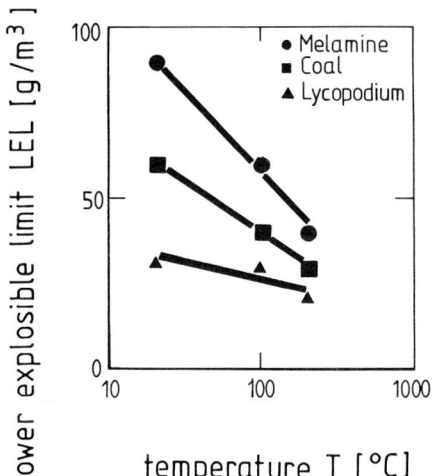

Fig. 56. Influence of temperature upon the lower explosible limit (LEL)

An increase of temperature will reduce the value of LEL (Fig. 56). This is more pronounced the higher the value at ambient temperature.

4.3.1.5 Explosion Pressure Versus Explosion Violence

The dust testing procedure described in the following section, was developed in 1966 [21] and meets all requirements of Sect. 4.3.1.1. The apparatus in use has been proven as a test instrument [55] and is internationally accepted as a standard [56].

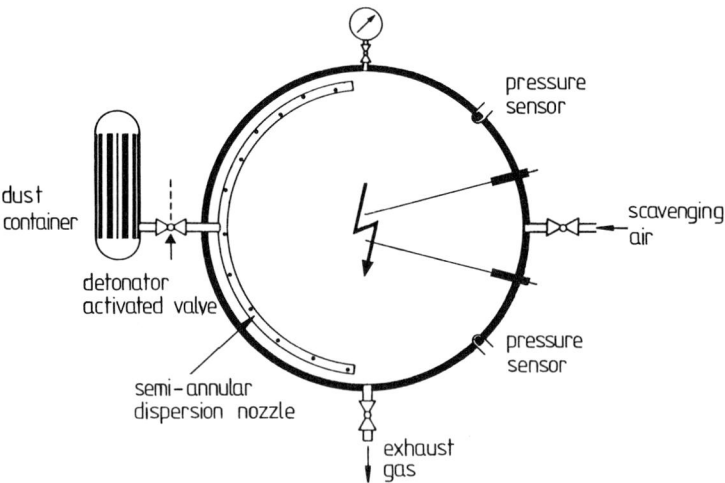

Fig. 57. Test apparatus ($V = 1\text{-m}^3$) for the determination of the explosion characteristics of combustible dusts (schematic)

4.3.1 Combustible Dusts

A pressure-resistant vessel with a volume of V = 1-m³ (Figs. 23 and 57) is used as an explosion chamber. Its length corresponds approximately to its diameter. Before testing, the dust is weighed into a 5.4-l dust container equipped with a fast-acting valve. After the valve opens, the dust is dispersed into the explosion chamber through a semiannular, perforated half-ring with 13 holes of 6-mm diameter each. Then, after a well-defined "ignition delay time" (t_d), the dust cloud is ignited by a suitable ignition source mounted in the center of the explosion vessel (see sect. 2.3. Pyrotechnical Igniters: total energy content E = 10 kJ).

The time-pressure behavior of the dust explosion in the 1-m³ vessel is monitored with pressure sensors (normally piezoelectrical pressure sensors) and recorded with luminous points.

Figure 58 shows the 1-m³ test vessel.

This dust testing procedure creates fairly homogeneous dust/air mixtures and ensures that the course of test explosions is reproducible.

For dust explosions in closed vessels, the rate of pressure rise dp/dt (Fig. 59) at any dust concentration is a measure of the explosion violence. The peak values which occur with the explosion of a combustible dust are:

– maximum explosion pressure p_{max}
– maximum rate of pressure rise $(dp/dt)_{max}$

These peak values are obtained from tests over a wide range of concentrations (Fig. 60). The two peak values normally occur at different dust concentrations.

Fig. 58. 1-m³ test vessel, ready for dust testing

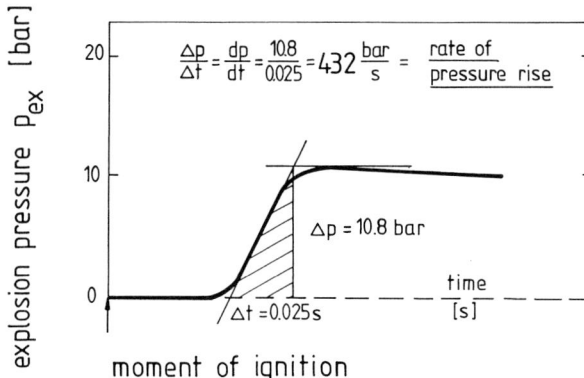

Fig. 59. Definition of the rate of pressure rise of a dust explosion (any concentration)

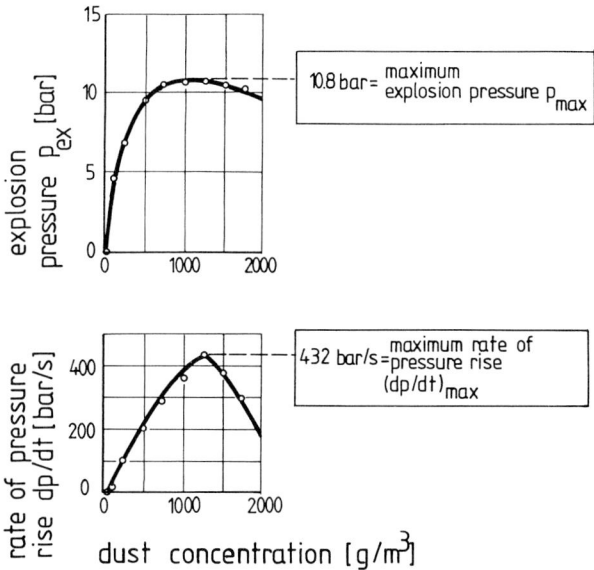

Fig. 60. Determination of the peak explosion characteristics of combustible dusts

The turbulence of the dust/air mixture at the moment of ignition exerts a significant influence on the above-mentioned explosion characteristics. Figure 61 shows this influence for two dusts. This degree of turbulence is assigned a special significance for the maximum rate of pressure rise $(dp/dt)_{max}$. With decreasing delay time t_d (increasing turbulence) the maximum rate of pressure rise increases continuously. Therefore, for dust testing in the 1-m³ explosion vessel a constant ignition

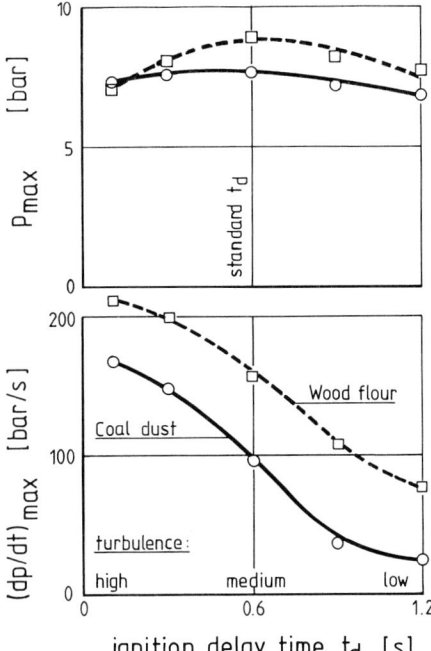

Fig. 61. Influence of the ignition delay time t_d (turbulence) upon the explosion characteristics of combustible dusts

delay time of $t_d = 0.6$ s has been selected. This value is only valid if dust containers of 5.4-l are used. The resulting time span ensures complete injection of the contents, i.e., the dust concentration in the air within the explosion chamber is known.

If the control of the standard delay time t_d is not maintained, the actual dust concentration in the 1-m³ chamber will vary. Either the dust/air mixture will not yet be dispersed ($t_d < 0.6$ s) or the process of sedimentation will have started already ($t_d > 0.6$ s).

For the determination of the maximum explosion characteristics over a large dust concentration range it is sometimes unwise to use the standardized 5.4-l dust container – expecially for dusts with a low density. It may become necessary to use 10-l dust containers. In this case, the ignition delay time t_d must be changed in such a way that the results determined with the 10-l container are the same as the ones obtained with a 5.4-l container.

Figure 62 shows that the larger dust container (with the same propellant pressure of $p_T = 20$ bar, air) has produced a higher turbulence for the dust/air mixtures, compared with the smaller ones. The maximum rate of pressure rise $(dp/dt)_{max}$ is more influenced than the maximum explosion pressure (p_{max}). Therefore, in order to obtain the same results with the 10-l container as with the 5.4-l container, an ignition delay time of $t_d = 0.9$ s must be used (Fig. 63).

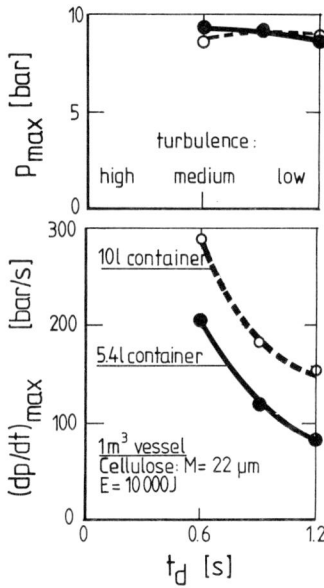

Fig. 62. Influence of the container volume and the ignition delay time upon the explosion characteristics of a dust

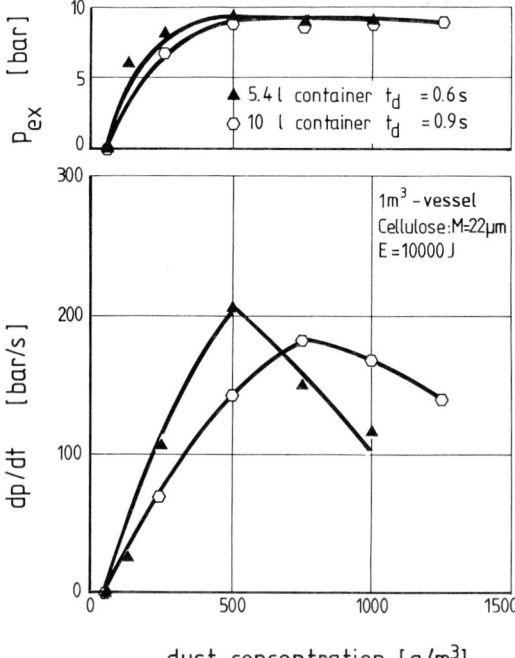

Fig. 63. Correlation between the dust concentration and the explosion characteristics using different dust containers and ignition delay times

4.3.1 Combustible Dusts

If the test volume is increased from 1-m³ to larger closed test vessels, the maximum explosion pressure p_{max} will remain constant. However, the maximum rate of pressure rise $(dp/dt)_{max}$ decreases according to the "Cubic Law":

$$(dp/dt)_{max,1} \cdot (V_1)^{1/3} = (dp/dt)_{max,2} \cdot (V_2)^{1/3} = \text{const} = K_{St}$$

where K_{St} represents a test- and procedure-specific characteristic value with the dimensions bar·m/s; it is independent of the test volume.

Figure 64 compares the explosion characteristics measured in vessels of different volumes with those measured in the standardized 1-m³ vessel.

V = 1-m³; one 5.4-l dust container
V = 2.4-m³; two 5.4-l dust containers
V = 10-m³; ten 5.4-l dust containers

The figure shows that the values for the maximum explosion pressure p_{max} are independent of the volume of the explosion vessel within the range of accuracy. Only the shape of the vessel can influence the peak value. In spherical vessels, higher pressures are observed, compared with cyclindrically shaped vessels.

The differences in the height-to-diameter ratios of the various vessels are as follows: for the 1-m³- and 10-m³-vessel, the ratio is 1; for the 2.4-m³ it is 1.55. This explains the slightly lower results obtained with the 2.4-m³ vessel.

Figure 64 shows also that the K_{St}-value for different dusts is independent of the volume of the explosion vessel.

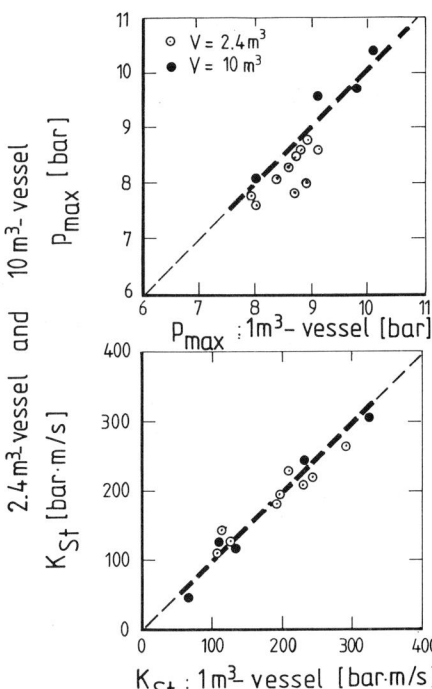

Fig. 64. Comparison of explosion characteristics of combustible dusts in vessels of different volumes

Therefore, the validity of the "Cubic Law" is proven for the course of explosions of combustible dusts in closed vessels in this size range. The above-described agreement of the peak values (p_{max} and K_{St}) is only valid if the dust testing procedure is correctly followed. This means that the number of 5.4-l dust containers must be proportional to the volume of the explosion chamber (see description of Fig. 64). Also the dust must be distributed from all dust containers through the usual perforated half-ring, and the source of ignition must consist of two pyrotechnical ignitors with a total energy of E = 10 kJ. The ignition must be activated after an ignition delay time $t_d = 0.6$ s. A corresponding number of 10-l dust containers is also acceptable provided that the ignition delay time is set at $t_d = 0.9$ s (Figs. 62/63).

Every deviation from the above-mentioned condition (e.g., changed volume of the dust storage container, changed propellent pressure, changed discharge valve of the storage container, change in the perforated half-ring for the dust distribution) can influence the explosion characteristics – especially the K_{St}-value. The results of such changes can be so drastic that even if the standard ignition delay time is used, there will be no correlation with the results from the standard 1-m³ vessel. If such modified systems for dust investigations have to be used, the ignition delay time has to be selected in such a way that the same results from testing a large number of dusts are obtained (within the range of accuracy) as are obtained from a 1-m³ vessel [55, 56].

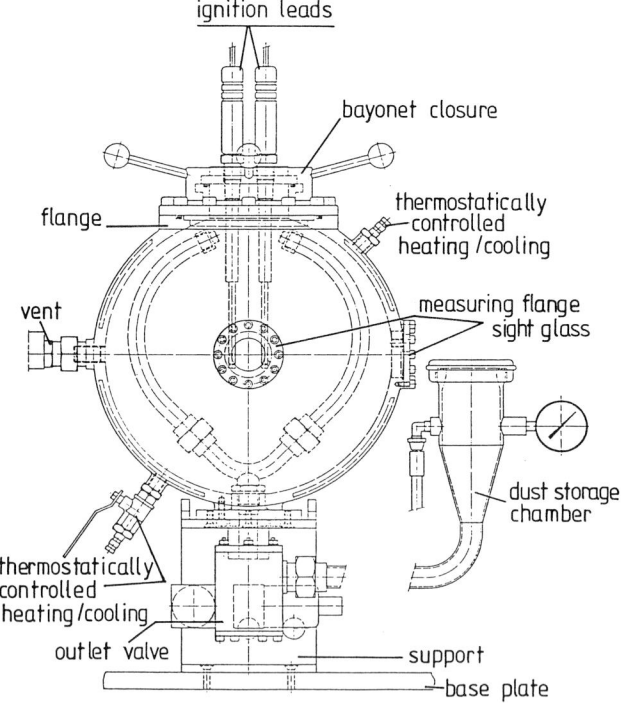

Fig. 65. (Legend see p. 63)

4.3.1 Combustible Dusts   63

The determination of explosion characteristics in large vessels involves a great deal of effort. Therefore, the question arose concerning what "minimum volume" could be used for which the "Cubic Law" remains valid.

Siwek [57–59] systematically investigated a large number of combustible dusts in spherically shaped equipment having varying volumes with correspondingly balanced dust storage containers. He recognized that a connection exists between the surface-to-volume ratio of the laboratory apparatus used and the values of the explosion characteristics. Siwek also determined the "minimum volume" for dust testing to be V = 20-l. This knowledge was used to establish a laboratory apparatus (Fig. 65) which works in principle according to the same procedure used for the standard 1-m^3 vessel (Fig. 66). The dust storage container for the 20-l apparatus was reduced to 600 ml.

Again, the dust dispersion occurs through a perforated half-ring, now with 112 holes of 3 mm diameter each. Because the small dust storage chamber empties quickly, the ignition delay time between the beginning of the dust dispersion and the activation of the ignition source is only t_d = 0.06 s. Prior to each test, the 20-l apparatus must be evacuated to an absolute pressure of 400 mbar. This is necessary to ensure that the dust explosion will be initiated at normal atmospheric pressure

b

Fig. 65 a/b. 20-l laboratory apparatus for the determination of the explosion characteristics of combustible dusts (**a**) schematic, **b**) actual apparatus)

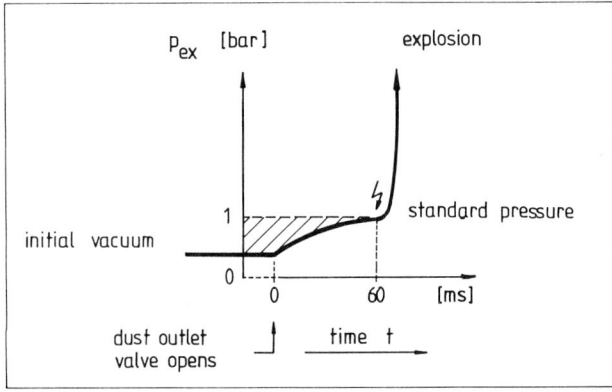

Fig. 66. 20-l laboratory apparatus: schematic presentation of the investigation procedure

(1 bar absolute) after expansion of the previously compressed air into the storage chamber.

Figure 67 shows the explosion characteristics based on a large number of dusts, measured in the 20-l laboratory apparatus and in the 1-m³ vessel, using the same ignition source (pyrotechnical ignitors with a total energy E = 10 kJ). The results obtained in the 1-m³ vessel are based on a starting initial pressure of $p_i = 0.1$ bar (1.1 bar absolute). This is inherent to the dust investigation test procedure. For the above-mentioned correlation (Fig. 67), these results have to be corrected to a pressure of 1.0 bar (absolute).

Based on Fig. 67, the following conclusions can be drawn:

- the values for the maximum explosion pressure p_{max} in the laboratory apparatus are slightly lower than the ones measured in the large test apparatus (1-m³ vessel), but they can be adjusted by a correlation graph,

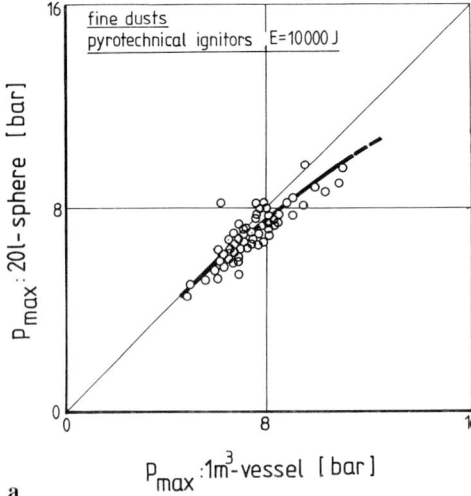

a

Fig. 67 a. (Legend see p. 65)

4.3.1 Combustible Dusts

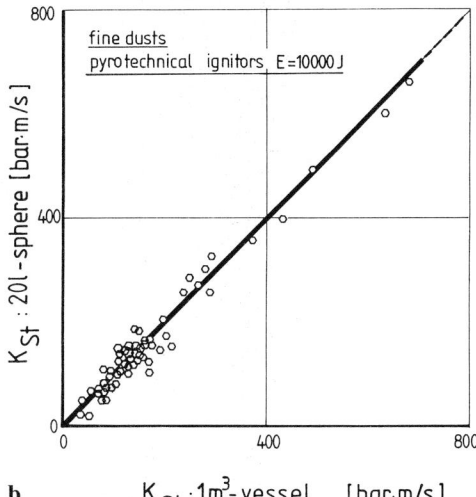

Fig. 67 a/b. Comparison of the explosion characteristics of combustible dusts measured in the 20-l laboratory apparatus and in the 1-m³ vessel; $p_i = 1.0$ bar (absolute)

- the K_{St}-values are the same in both apparatus up to highly reactive metal dusts ($K_{St} = 700$ bar·m/s).

Figure 68 shows that this is also valid for various coating powders, i.e., combustible dusts having the same process origins but not the same median values. In this case, the measured maximum explosion pressures from the laboratory sphere have been corrected.

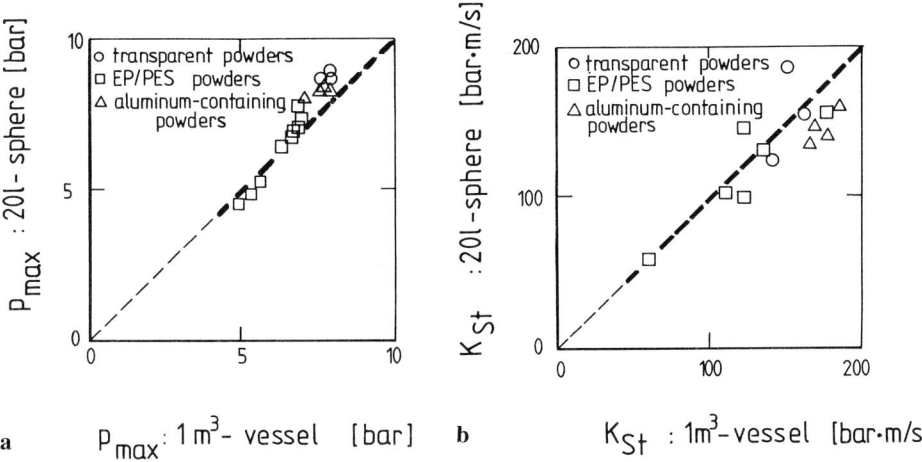

Fig. 68 a/b. Explosion characteristics of coating powders (comparison of the values: 20-l laboratory apparatus/1-m³ vessel)

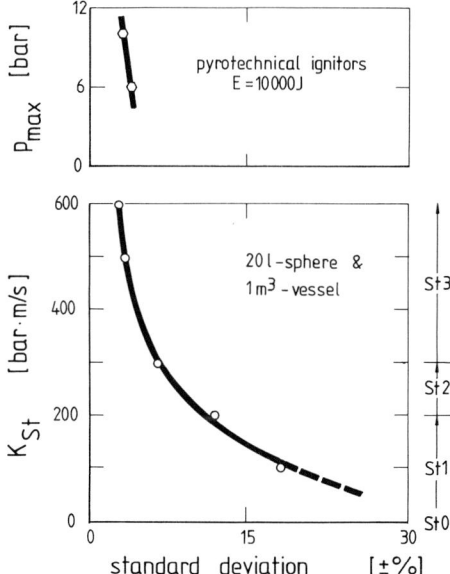

Fig. 69. Standard deviation of the explosion characteristics for combustible dusts

Figure 69 shows the standard deviations of both explosion characteristics for the laboratory and the larger apparatus. Again the powerful standard ignition source was used. The maximum explosion pressure can be determined with high accuracy, independent of the explosion velocity. The accuracy of the dust-specific value K_{St} shows a marked decrease towards lower values. In absolute values, the accuracy is practically the same.

For the investigation of the explosion behavior of combustible dusts at elevated temperatures, an electrically heated 20-l laboratory sphere (Fig. 70) was developed, which allows explosion testing at temperatures up to 250 °C. The test procedure is the same as shown in Fig. 66. Prior to testing, the pre-evacuation pressure must be adjusted according to the test temperature.

For example:

– at a test temperature of 100 °C: $p_i = 380$ mbar and
– at a test temperature of 250 °C: $p_i = 250$ mbar

With this procedure, initiation of the dust explosion occurs at normal atmospheric pressure independent of the test temperature.

For further investigations, an apparatus having a volume $V = 38$-l (Fig. 71) was developed. It also works according to the procedure shown in Fig. 66. This apparatus is equipped with a dust storage chamber having a volume $V = 0.9$-l. The perforated half-ring for the dust injection utilized in the 38-l apparatus is the same one as in the 20-l apparatus. To obtain the same explosion characteristics as those measured in the standard 1-m³ vessel, it is necessary to use an ignition delay time of $t_d = 0.08$ s. The dust explosion is initiated by two pyrotechnical ignitors having a total energy $E = 10$ kJ.

4.3.1 Combustible Dusts 67

Fig. 70. Electrically heated 20-l laboratory apparatus for the determination of the explosion characteristics of combustible dusts at elevated temperature

Fig. 71. 38-l apparatus for dust investigations

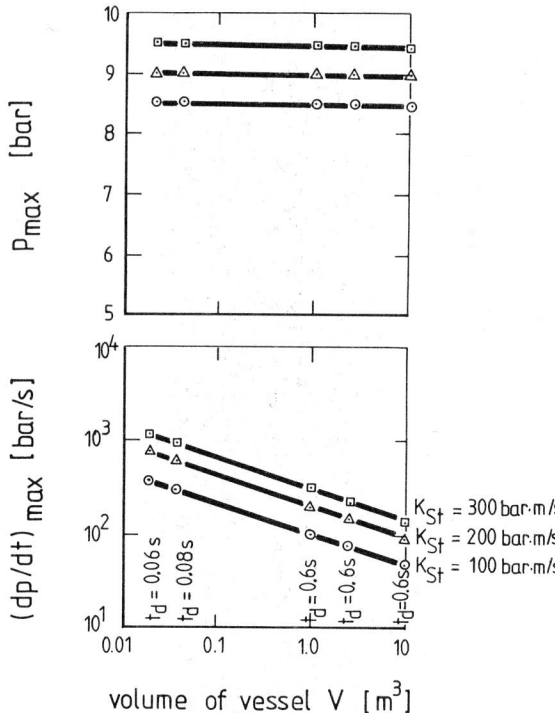

Fig. 72. Influence of vessel volume upon the explosion characteristics of combustible dust

Thus a series of dust-testing vessels of $V = 0.02$ to 10.0-m^3 is today available for dust investigations in closed equipment. They are used in connection with a prescribed test procedure and an exactly determined ignition delay time t_d (Fig. 72).

The maximum explosion pressure p_{max} measured in the above range of apparatus is independent of the volume of the vessels. The maximum rate of pressure rise $(dp/dt)_{max}$ decreases according to the "Cubic Law". Therefore, the K_{St}-value is a volume-independent, dust-specific characteristic.

The optimum concentration of the above-mentioned explosion characteristics is two to three times higher than the stoichiometric composition of the dust/air mixtures, which is in the range between 100 g/m^3 and 300 g/m^3 [53]. After the dust explosion, unburned or charred residues are often found.

The discussion up to now has involved closed vessels completely filled with explosible dust/air mixtures. If the vessels are only partially filled with the optimum dust concentration (Fig. 73) the maximum explosion pressure decreases almost linearly, while the K_{St}-value decreases exactly linearly with the "filling ratio".

The particle size distribution of a dust has an important influence on the explosion characteristics. Figure 74 shows this influence on four arbitrarily chosen dust types.

4.3.1 Combustible Dusts   69

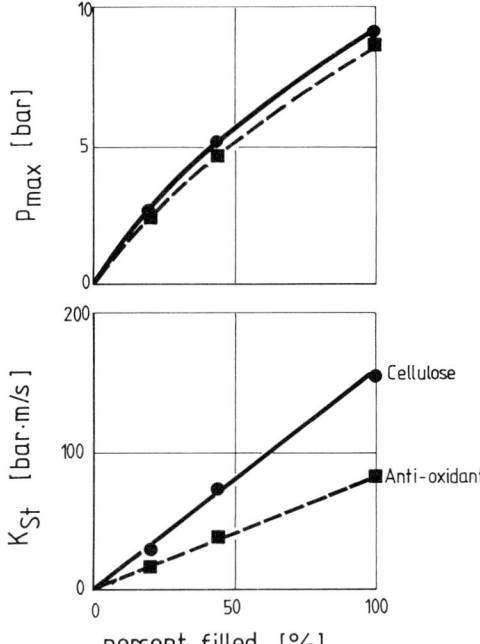

Fig. 73. Influence of the "filling ratio" upon the explosion characteristics of combustible dusts in a 7-m³ closed vessel

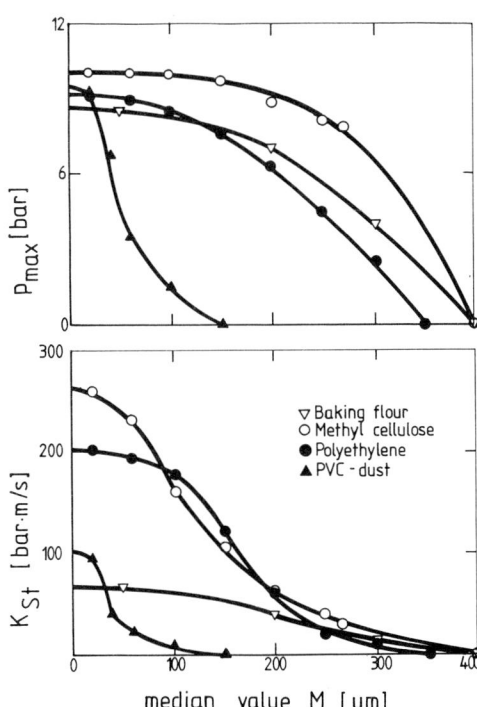

Fig. 74. Influence of the median value M upon the explosion characteristics of combustible dusts

Fine dusts react more violently than coarse dusts. The median value M therefore has a marked influence on the explosion violence (the K_{St}-value) and a less pronounced one on the maximum explosion pressure. Therefore, to obtain the maximum values for industrial applications, the dust samples used for the determination of the maximum explosion characteristics should have a median value $M \leq 63$ μm.

It also follows from Fig. 74 that, in general, dusts with a particle size of 400 to 500 μm cannot be caused to explode, even with powerful ignition sources.

It must also be remembered that fine dust can be created by abrasion of a portion of a coarse dust. If that is the case (Fig. 75), it has to be recognized that – depending on whether this particle size limit is exceeded considerably (methylcellulose) or only slightly (polyethylene) – an admixture of fine dust of 10% in the case of methylcellulose or 5% in polyethylene will be sufficient to make the mixture explosible. In fact, a relatively high maximum explosion pressure is reached. There is no clearly defined lower level of concentration for fine dust within coarse dust below which no protective measures have to be considered [23]. But an explosion hazard is always present if the content of the fine dust in the coarse, non-explosible material exceeds its lower explosible limit.

If combustible dusts with particle sizes between 400 and 500 μm are produced or handled, an explosion hazard can still exist if sufficient fine dust is produced, e.g., through abrasion, creating an explosible mixture.

Some dusts, e.g., red phosphorus, can ignite themselves by simply exiting from the ring nozzle. The explosion occurs before the induced ignition and therefore at an ignition delay time below the standard one. A determination of the explosion characteristics according to the test procedure is therefore not possible without difficulties. It is known that the self-ignition of phosphorus decreases by lowering the

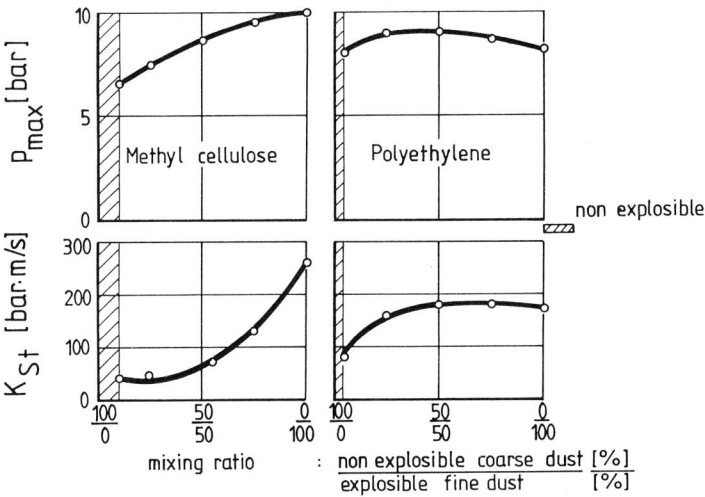

Fig. 75. Explosion characteristics of mixtures of coarse and fine dusts

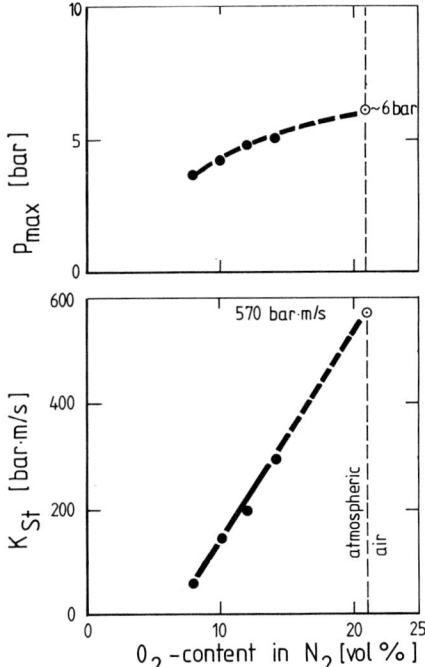

Fig. 76. Determination of the explosion characteristics of red phosphorus in oxygen-lean atmospheres (20-l laboratory apparatus; E = 10 kJ)

oxygen concentration. In the process of determining the explosion data in the 20-l sphere at a low oxygen level, a relationship between the explosion values and the oxygen concentration in nitrogen has been found (Fig. 76). This correlation is only valid if the oxygen content in nitrogen is not too low. It is possible, according to Fig. 76, to extrapolate the explosion characteristics to normal atmospheric pressure. Control measurements in normal air without self-ignition proved the results gained through the extrapolation procedure.

Often the statement is made that dusts containing a few percent of water can no longer generate explosive dust/air mixtures. Figure 77 contradicts this assumption. A decrease of the explosion violence is normally observed at relatively high water content. However, many products exceeding 10% water content cannot be dispersed (whirled up) properly. Therefore, the hazard decreases for the formation of explosible dust/air mixtures from dust deposits. Products containing solvents show a significantly different behavior. Figure 78 shows this influence of different solvents on two pharmaceutical dusts.

In case of the product containing ethanol, the explosion characteristics will first be decreased by increasing the solvent concentration. When a high solvent concentration is reached, the explosion characteristics begin to increase. The reason is that at sufficient solvent concentration the product agglomerates and cannot be properly dispersed. Also at still higher ethanol concentrations in the explosion chamber, an ethanol/air mixture will be developed and ignited through the ignition

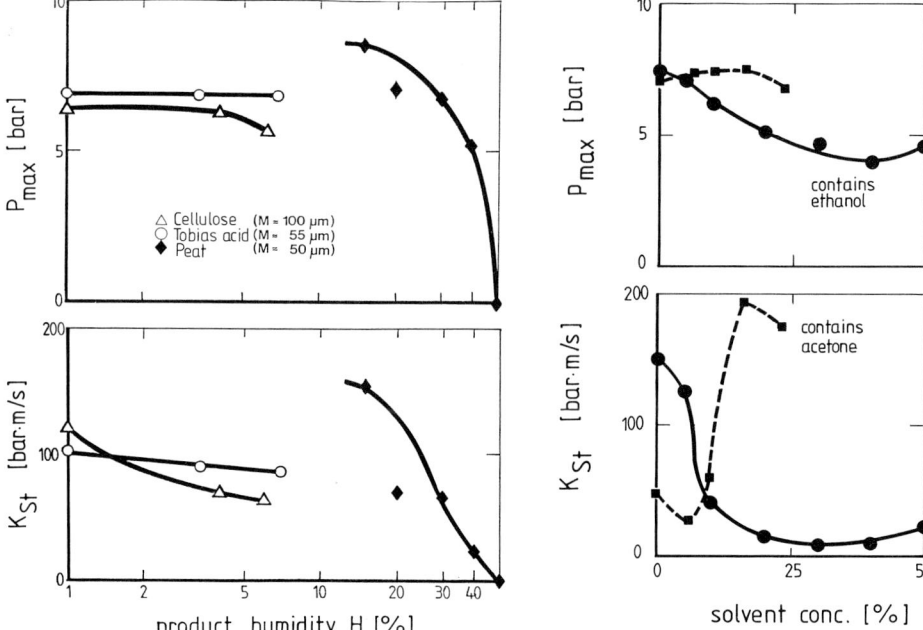

Fig. 77. Products containing water: influence of product humidity upon the explosion characteristics

Fig. 78. Pharmaceutical products containing solvents: influence of product humidity upon the explosion characteristics

source instead of the dust product. The observed effects are different for dusts containing acetone. Acetone evaporates four times faster than ethanol. Therefore, the values for the explosion characteristics show that at acetone concentrations greater than 15% the stoichiometric combustion mixture in the explosion chamber is exceeded.

The above finding is of great technical safety importance concerning the assessment of risks created by the processing of products containing solvents, e.g., in fluidized bed driers.

The course of an explosion of dispersed combustible dust in closed equipment is also influenced by the initial pressure p_i, the pressure existing at the moment of ignition. Figure 79 shows, with the example of three combustible dusts, that the maximum explosion pressure p_{max} and the K_{St}-value are proportional to the initial pressure p_i. Also, the optimum dust concentration for the explosion characteristics increases proportionally as the initial pressure is increased.

A doubling of the initial pressure p_i causes a doubling of the explosion characteristics, and it must be stated again that the maximum explosion values at these elevated initial pressures can only be obtained at relatively high dust concentrations. Because the straight lines do not intersect the origin of the coordinates, below a

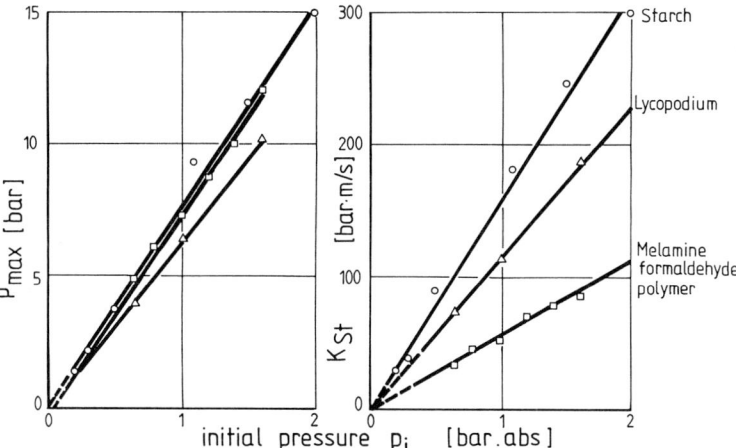

Fig. 79. Influence of the initial pressure p_i upon the explosion characteristics

limiting pressure of about some 10 mbar (absolute), explosions of dust are apparently no longer possible.

In industrial practice, combustible dusts are often produced, manufactured, and transported at elevated temperatures; so it is important to know the influence of the temperature on the course of the explosion characteristics [60]. Figure 80 shows the influence of the temperature on the maximum explosion pressure at normal pressure.

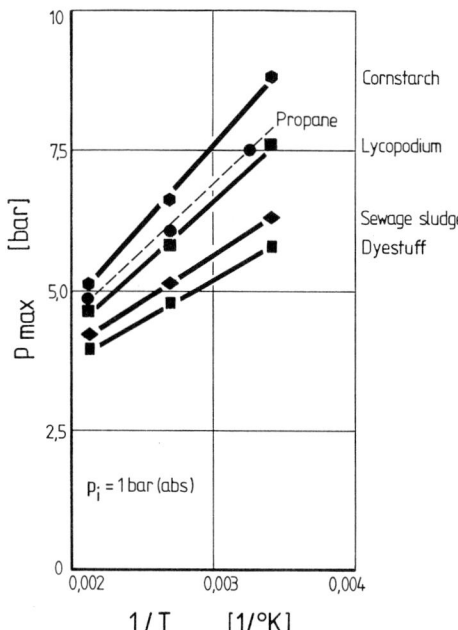

Fig. 80. Influence of the temperature upon the maximum explosion pressure (initial pressure = normal pressure)

The explosion pressure decreases linearly with the reciprocal value of the increasing temperature. The same trend was found with combustible gases. The explosion pressure varies the same way if the initial pressure p_i is varied (Fig. 81).

The observed variation of the maximum explosion pressure with temperature is due to variations in the oxygen concentration (in nitrogen) which also changes in accordance with the ideal gas law. Figure 82 shows an example with two dusts. The

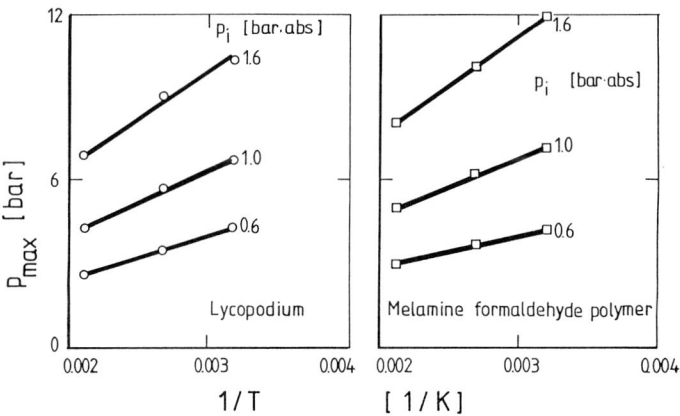

Fig. 81. Influence of the temperature T and the initial pressure p_i upon the maximum explosion pressure p_{max}

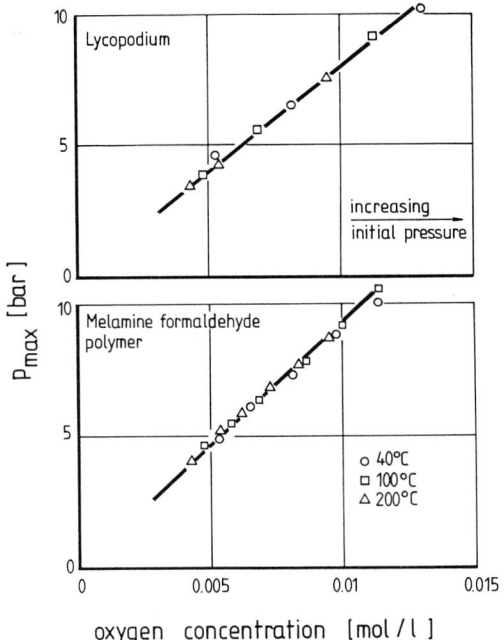

Fig. 82. Influence of the oxygen concentration in nitrogen upon the maximum explosion pressure

oxygen concentration is the principle variable affecting the maximum explosion pressure in a closed vessel. Changes in temperature or initial pressure have less effect.

This is of great importance in industry. A constructive protection measure called "explosion venting" is often used on spray driers. Such driers have large volumes and it is often difficult to accommodate the required venting areas. However, two factors may allow the reduction of the size of the vent area in comparison with that suggested by the actual guideline [23]. First, the area requirement is less at low maximum explosion pressures. And secondly, the relatively high drying temperature lowers the explosion pressure (Fig. 80).

Figure 83 shows the behavior of the K_{St}-value for these dusts at elevated temperatures. Again, there is a linear relationship between the oxygen concentration in nitrogen and the dust-specific characteristics. It is well known that a reduction of the oxygen concentration at ambient temperature decreases the maximum rate of pressure rise and therefore the K_{St}-value. As can be seen, there are dusts (e.g., Lycopodium) which also follow this law at elevated temperatures. For other dusts (e.g., pea flour), a temperature increase does not influence the rate of the explosion (within the accuracy of the test results). Evidently, the increased temperature accelerates the explosion reaction sufficiently to compensate for the reduction of the K_{St}-value expected from the lower oxygen content. In the case of certain dusts (e.g., dyestuff C), an increase in temperature causes a linear increase in the K_{St}-value. The course of the explosion reaction is accelerated sufficiently to more than compensate for the impeding effect of the reduced oxygen content.

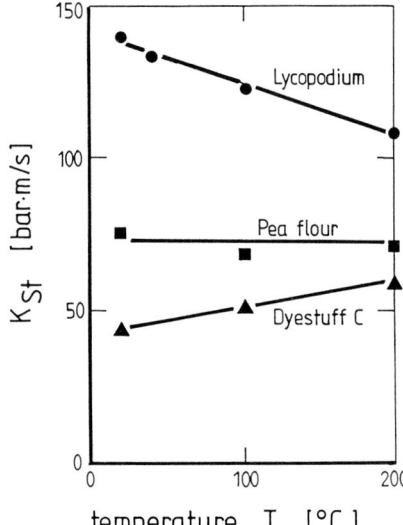

Fig. 83. Influence of the oxygen concentration in nitrogen upon the K_{St}-value (initial pressure = normal pressure)

In general, a temperature increase up to 300 °C has a minor influence upon the K_{St}-value and is relatively unimportant for practical applications. If the initial pressure is changed in addition to the temperature (Fig. 84), then Lycopodium will be subjected to an influence similar to the one shown for the maximum explosion pressure (Fig. 82). The oxygen content in nitrogen solely determines the level of the K_{St}-value. However, in the case of the polymer melamine, the temperature as well as the initial pressure influences the explosion behavior.

Based on current knowledge, it seems that the behavior of the dust-specific characteristics of lycopodium is representative of relatively violently reacting, easily ignitable or normally ignitable dusts. Dusts which react relatively weakly and are hard to ignite, however, show a behavior similar to the K_{St}-value of melamine.

As already mentioned, a dust explosion in the described test vessels is initiated by a strong ignition source (pyrotechnical ignitors with a total energy content E = 10 kJ). Such ignition sources are highly unlikely in industrial settings. But on the other hand, high energies are necessary for testing in order to detect all combustible dusts which are potentially explosible in industrial practice. Therefore, the question is raised whether the high energy level used in dust testing influences the explosion characteristics of combustible dusts which describe the course of an explosion.

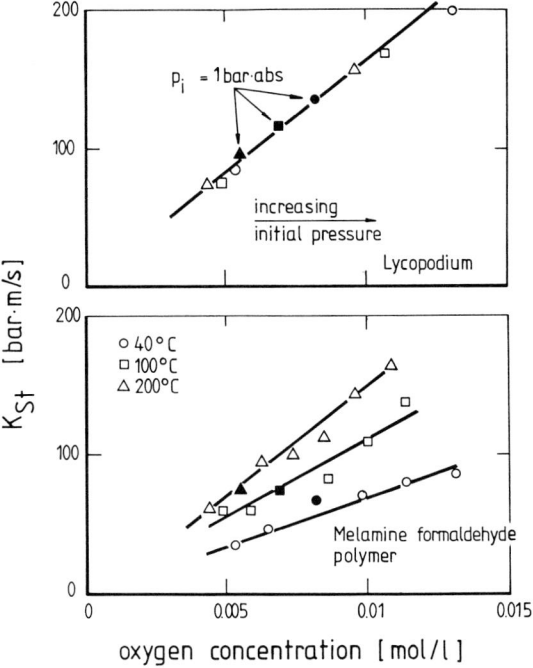

Fig. 84. Influence of the oxygen concentration in nitrogen upon the K_{St}-value (initial pressure variable)

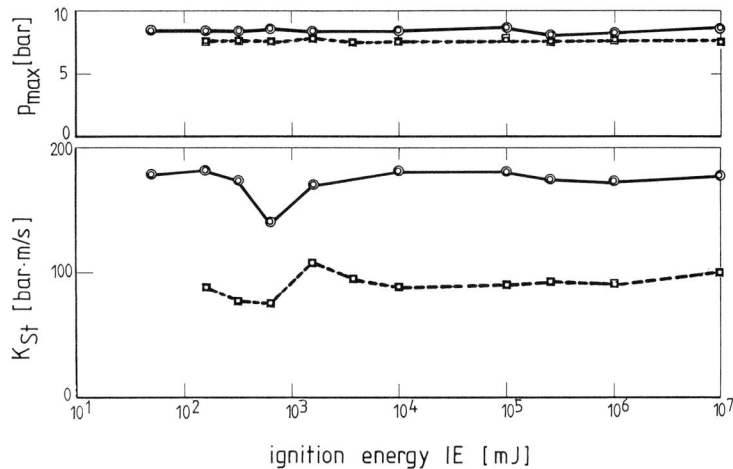

Fig. 85. Influence of the ignition energy upon the explosion characteristics of easily or normally ignitable dusts

Figure 85 is proof that this is not the case. The dusts shown here are easily or normally ignitable by condenser discharge. For these dusts, the ignition source is not important. Weak condenser discharges as well as strong pyrotechnical ignitors, such as the ones specified for dust testing, give the same results.

There exists another group of dusts, which are difficult to ignite (Fig. 86). A decrease in the ignition energy will normally result in a linear reduction of the K_{St}-value. The maximum explosion pressure is generally not affected by such an influence. Only with dusts which are very difficult to ignite (e.g., dyestuff), is this decreasing tendency noticeable.

These conclusions are valid for all test apparatus having a volume $V = 0.02$ to 1.0-m^3 and which are operated in accordance with the prescribed procedure for dust testing (Fig. 72).

With the large number of dusts produced and handled in industrial operations it was appropriate to classify them in accordance with their violence of explosion as expressed by the K_{St}-value. Such classes are used in the design of constructive safety measures against dust explosions. The range of the K_{St}-values covered by each class is defined in Table 9.

It must be made clear that a given dust explosion class is only an indication of the expected course of a dust explosion. The classification gives no indication as to how easily a given dust ignites. Relatively weak reacting dusts may ignite very easily, while violently reacting dusts may be very difficult to ignite with a condenser discharge as the ignition source. Therefore, the classification of a dust into a given dust explosion class does not indicate or address the probability of an explosion occurring nor does it predict the consequences from such an explosion in industrial practice.

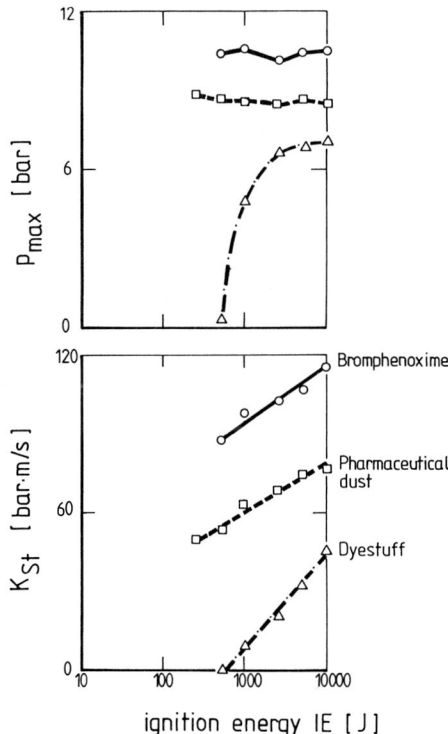

Fig. 86. Influence of the ignition energy upon the explosion characteristics of dusts which are difficult to ignite

Table 9. Correlation between K_{St}-values and dust explosion class (E = 10,000 J)

K_{St} [bar · m/s]	Dust explosion class
> 0–200	St 1
201–300	St 2
> 300	St 3

The explosion characteristics of more than 500 dusts have been systematically determined. Figure 87 shows the distribution or frequency per dust explosion class or K_{St}-value, respectively. The majority of the combustible dusts fall into the dust explosion class St 1.

Figure 88 shows the distribution of the maximum explosion pressure within the three dust explosion classes for more than 500 dusts. Each of the three dust explosion classes has an optimum pressure range which occurs most frequently. This optimum pressure range increases at higher dust explosion classes.

4.3.1 Combustible Dusts

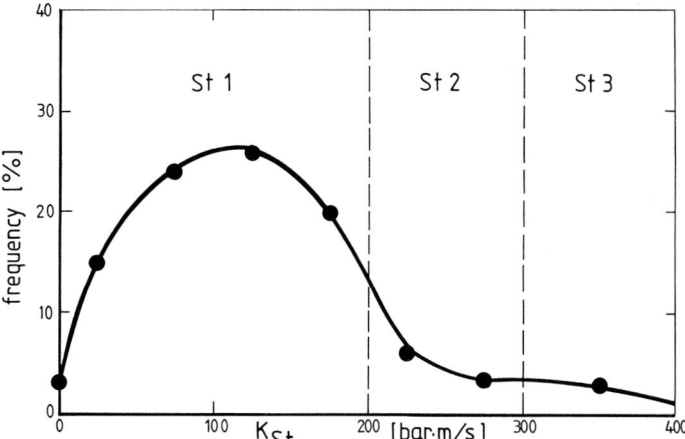

Fig. 87. Frequency of the dust explosion classes for more than 500 dusts

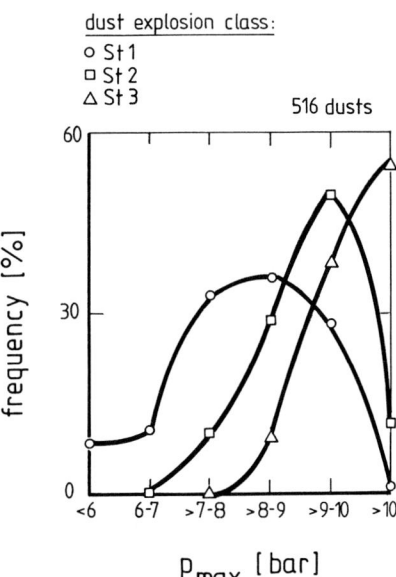

Fig. 88. Correlation of the maximum explosion pressure with dust explosion class

Table 10 summarizes the explosion characteristics and the dust explosion classes of some combustible dusts.

The values in Table 10 are for reference purposes only, because the configuration of the dust particles may influence the course of the explosion as explained earlier.

Additional explosion characteristics are listed in the original research paper [61]. Their usefulness is limited due to two main factors: 1) the various test methods used [1 m³-vessel (Fig. 57) versus modified Hartmann apparatus (Fig. 21)] and 2) especially the effects of particle size (Fig. 74). For example, flour dust (corn) did not ignite in the 1-m³ vessel due to its median particle size value of M = 550 μm. But the fine dust M ⩽ 63 μm) ignited in the modified Hartmann apparatus as expected. Such a fine dust would, of course, also ignite in the 1-m³ vessel.

Table 10. Explosion characteristics of combustible dusts (M <63 μm)

Type of dust	p_{max} [bar]	K_{St} [bar · m/s]	Dust explosion class
PVC	8.5	98	St 1
Milk powder	9.7	130	
Polyethylene	8.8	131	
Coal dust	8.2	135	
Metallic soap	8.4	140	
Chicory	8.5	157	
Toner	7.5	166	
Resin dust	8.9	174	
Brown coal	10.0	176	
Wood dusts	9.4	208	St 2
Cellulose	9.8	229	
Organic peroxide	9.0	273	
Bromphenoxime	11.9	342	St 3
Pigments	10.7	344	
Musk (nitrated)	10.0	360	
Red phosphorus	~ 6.0	~570	
Aluminum	12.5	650	

4.3.2 Flock

4.3.2.1 Preliminary Remarks

Flock is a monofilament fiber having a range of deniers of 0.8 to 20 with lengths of 0.5–2.5-mm (Fig. 27).

In 1968, the VDE-ordinance 0147 was validated. In this ordinance the ignition and explosion behaviors of coating powders and flock were treated equally. This triggered systematical testing [62, 63].

The following statements are based on flock made of polyamide 66, polyamide 6, acrylics, nylon, polyesters, and cotton with deniers 0.5–6.0 and lengths of 0.5–1.5-mm. The flock was available in cut or ground form. Various preparations have also been tested.

Due to the high bulk density of flock, it was necessary to use the large 1-m^3 vessel for the explosion tests. The dispersion of the flock with the standard semi-annular nozzle (Fig. 57) was quite difficult because of plugging. Therefore, it was necessary to develop a special nozzle (Fig. 89) for proper dispersion. This is permissible because the explosion characteristics of combustible dusts remained the same, within the range of accuracy, as the ones obtained using the so-called semi-annular nozzle (Fig. 90).

The test results indicate that the product of denier times length of cut (weight of flock particle) determines the explosion behavior and not the type of flock. This product has the same significance for flock as the median particle size M for combustible dusts.

Fig. 89. Special nozzle for the dispersion of flock into the large vessel (V = 1-m^3)

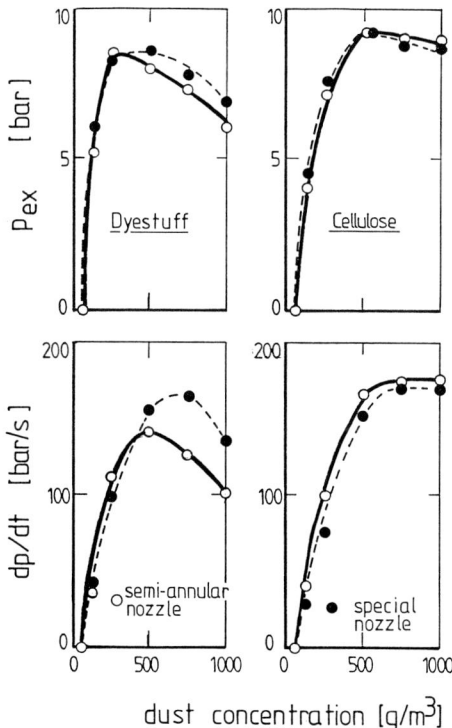

Fig. 90. Comparison of explosion indices obtained with different dust dispersions in the large vessel ($V = 1\text{-m}^3$)

4.3.2.2 Explosible Limits

The lower explosible limit of flock depends upon the value of the product denier times length of cut, analogous to the median particle size of dusts (Fig. 91).

With increasing value of product denier times length of cut there is a marked increase in the lower explosible limit (Table 11). The figures in Table 11 are based upon the usual ignition energy of 10,000 J. Analogous to dusts, the question can be raised concerning to what extent the ignition energy influences the LEL of flock. Figure 92 shows that such influence is much more pronounced.

Acrylic flock (0.54 denier/0.9 mm; 1.17 denier/0.5 mm), for example, has a very low LEL of 15 g/m³ at a high ignition energy. With an ignition energy of just IE = 1 J the observed level increases to an LEL of 125 g/m³. Such an influence is more pronounced for flock with higher explosible limits. Based on this observation, the concentration in the flock machines should not exceed 125 g/m³ in order to prevent an explosion.

4.3.2 Flock

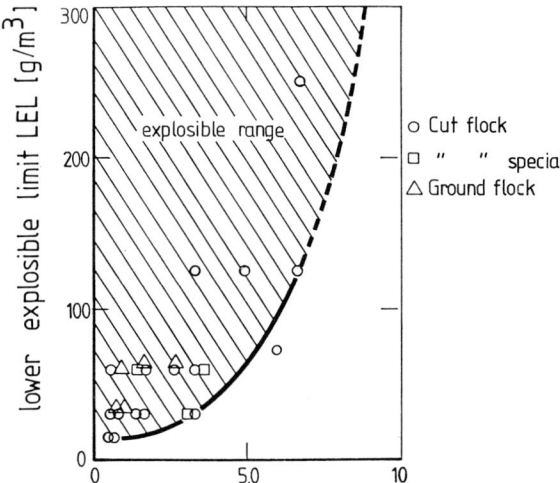

Fig. 91. Correlation of lower explosible limit of flock with the product denier times length of cut

Table 11. Correlation of lower explosible limit (LEL) of flock with the product: denier times cut length (E = 10,000 J)

1.11 denier × cut length in mm	LEL [g/m³]
0.54	15
3.30	30
4.90	60
6.70	125
8.50	250

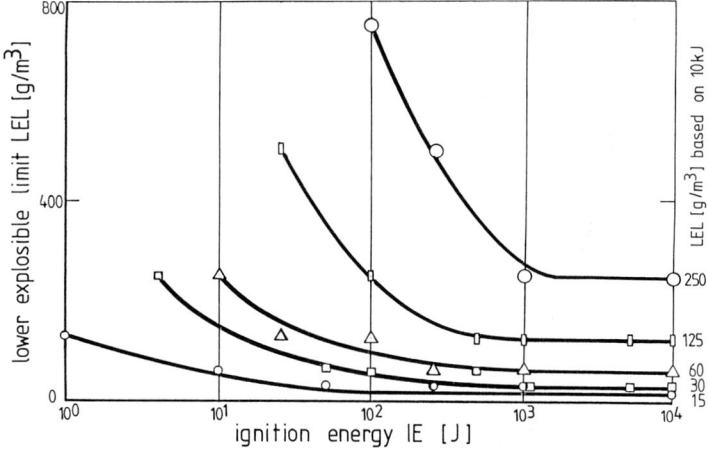

Fig. 92. Influence of ignition energy (IE) upon the lower explosible limit (LEL) of flock

4.3.2.3 Explosion Pressure / Violence of Explosion

The optimum explosion indices of flock have to be determined over a wide range of concentrations similar to requirements for combustible dusts. The results of such tests are shown in Fig. 93.

In following the limiting curve, it is obvious that the maximum explosion pressure and the K_{St}-value decrease with increased product denier times cut length. This behavior is similar to that observed with combustible dusts relative to the median particle size M (Fig. 74). The above figure also shows that there is no danger of an explosion for values of denier times cut length >9. In accordance with the definition of the dust explosion classes (Table 9), flock can be classified into the dust explosion class St 1 ($K_{St} \leq 200$ bar·m/s). The following correlation is given:

The optimum flock concentration which results in the maximum explosion pressure p_{max} and the K_{St}-value increases in general with increasing denier value (Fig. 94) from 250 g/m³ (0.54 denier) to 750 g/m³ (6 denier). This is true for both explosion characteristics.

The behavior of the optimum concentration of flock relative to the denier value is similar to the optimum concentration of combustible dusts relative to the median particle size M.

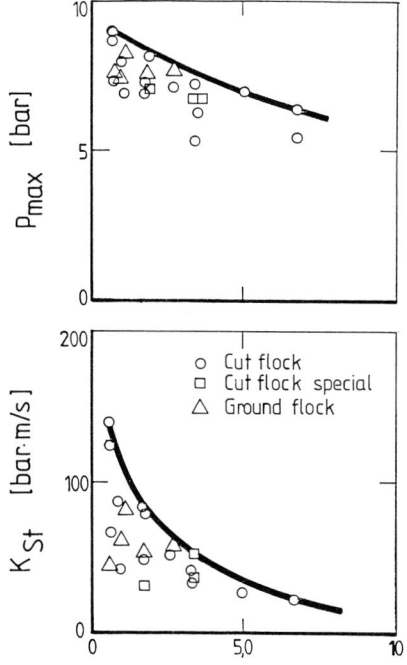

Fig. 93. Correlation of explosion indices with the product: denier times cut length

Table 12. Correlation of explosion indices with the product: denier times cut length (E = 10,000 J)

1.11 denier × cut length in mm	p_{max} [bar]	K_{St} [bar · m/s]
0.54	9.0	140
1.70	8.4	88
3.3	7.5	52
6.7	6.3	21

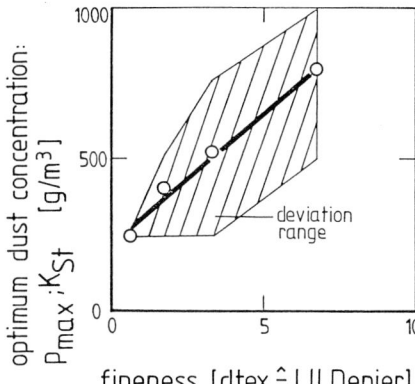

Fig. 94. Influence of optimum flock concentration upon explosion indices

Figure 95 shows the influence that the ignition energy has upon the explosion pressure p_{ex} and the pressure rise dp/dt. The values are not identical with the maximum pressure and the maximum pressure rise because they were not determined at optimum flock concentrations. The optimum explosion pressure is practically independent of the ignition energy. This is also true for the pressure rise if it is equal or larger than 60 bar/s. For less violently reacting flock, the pressure rise decreases with decreasing ignition energy. Here again, the behavior is similar to that of combustible dusts (Figs. 85 und 86).

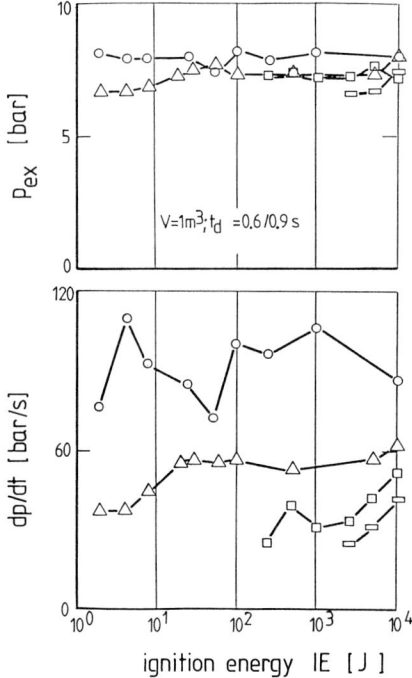

Fig. 95. Correlation of ignition energy with explosion pressure p_{ex} and pressure rise dp/dt

4.3.3 Hybrid Mixtures

4.3.3.1 Preliminary Remarks

Combustible dust and gases or combustible vapors may coexist in air during the processing and manufacturing of solvent-containing products or through carbonizing of overheated dusty products. The mixtures result from two sources, and therefore they are called *hybrid mixtures*. As mentioned earlier, Engler conducted the first experiments in 1885. Additional experiments were made pertaining to problems in the mining industry [21, 63]. It was determined that the violence of a coal dust explosion could be increased by adding methane to the air at a non-explosive concentration. Pellmont [7] systematically studied the explosion and ignition behavior of such hybrid mixtures, and recently additional findings were reported [62].

4.3.3.2 Explosible Limits

The behavior of the lower explosible limit (LEL) of hybrid mixtures is very important for practical safety audits in manufacturing plants. Figure 96 shows the correlation of the LEL of three combustible dusts with the content of propane in air. A linear decrease of the LEL of the solids can be noticed with increasing gas content.

A standardized presentation [64] is necessary in order to compare the lower explosible limits of different substances [21]. The stated concentration is normally referenced to the LEL of the pure component (Fig. 97). In general, there is a linear correlation of the LEL of hybrid mixtures following the law of Le Chatelier. Such a correlation has also been documented for hybrid mixtures of several components [65].

An exception to the rule exists in mixing PVC with methane, where the correlation is hyperbolic. This effect is probably caused by the methane because the same dust in a propane-air mixture follows the law of Le Chatelier. Methane has a lower normal velocity of combustion than propane. At present, no generally valid correlation can be given. However, newer test results confirm Engler's statement of 1885 that the admixture of non-explosible dust and gas-air mixtures will form an explosible hybrid mixture.

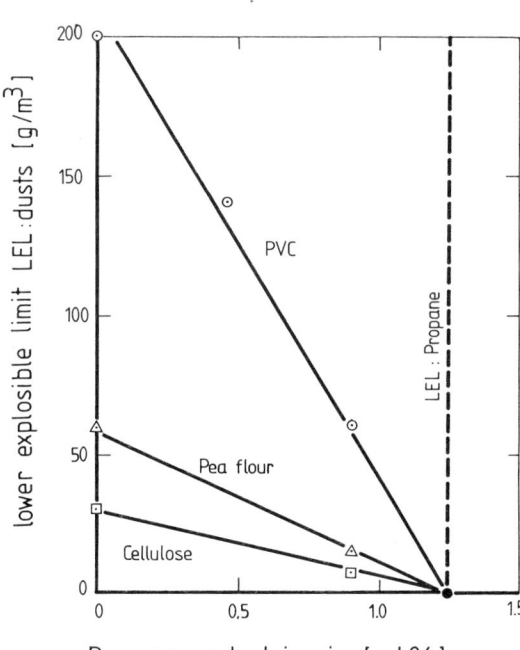

Fig. 96. Lower explosible limit of hybrid mixtures consisting of combustible dusts and propane

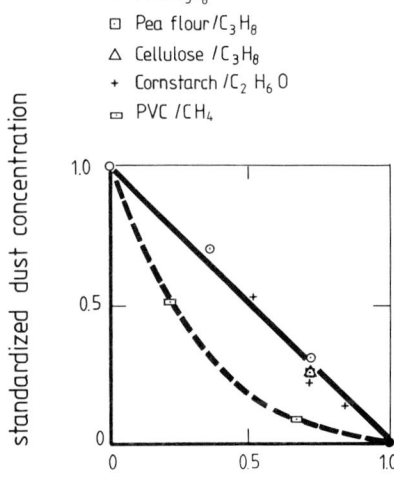

Fig. 97. Standardized presentation of the lower explosible limit of hybrid mixtures consisting of combustible dust with flammable gas

Furthermore, dusts which, for instance, will not ignite due to the particle size, even with a potent ignition source, will probably become explosible in an admixture of flammable gas or vapor.

In hybrid mixtures with a given dust concentration which is below the LEL, the required amount of flammable gas or vapor to reach an explosible mixture becomes smaller the lower the explosible limit.

4.3.3.3 Explosion Pressure/Violence of Explosion

Figure 98 shows the influence that an additional concentration of propane in air has on the explosion pressure and the pressure rise of cellulose.

An increasing propane concentration results in a continuous increase in indices, even if the propane-air mixture by itself is non-explosive (0.9 vol. %). It is still necessary to conduct such tests with the usual potent ignition source used for dusts.

Figure 99 depicts the correlation of propane concentration with the explosion indices of dusts. The addition of flammable gas has only a marginal influence on the maximum explosion pressure. The K_{St}-value, however, is influenced markedly. It increases linearly with increasing propane content and decreases only after the maximum gas concentration has been exceeded. In general, the dust-specific index for all three dusts is in the range of the K_G-value for the admixed flammable gas without the dust. (The K_G-value is the gas-specific index derived from the cubic

4.3.3 Hybrid Mixtures

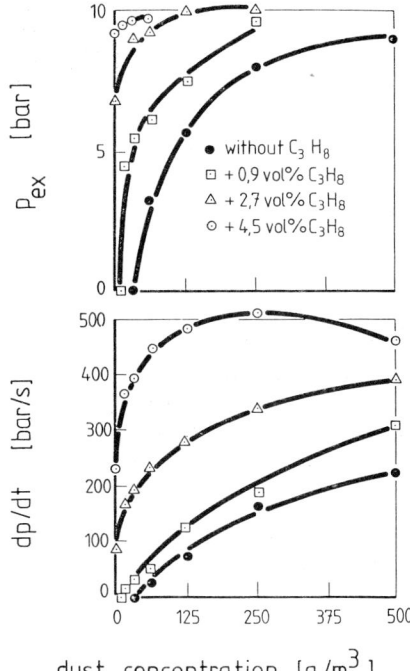

Fig. 98. Influence of the propane concentration in air upon the pressure rise of cellulose ($V = 1\text{-m}^3$; $E = 10,000$ J)

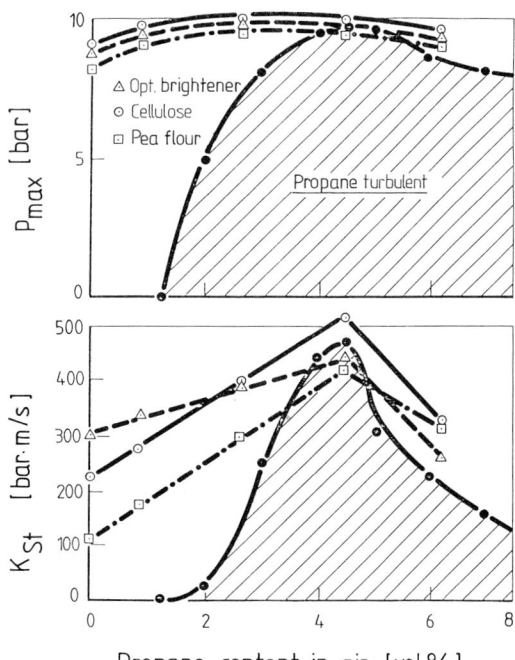

Fig. 99. Influence of the propane concentration in air upon the explosion indices of combustible dusts ($V = 1\text{-m}^3$; $E = 10,000$ J)

law). The mixing of a flammable gas with a dust has a lesser influence if the dust by itself reacts more violently (optical brightener). However, the influence is more pronounced with a weak reacting dust such as pea meal.

These findings are not only valid for large equipment but also for laboratory apparatus [63].

In designing protective measures against the outcome of dust explosions, one has to consider that the presence of flammable gases and vapors may classify the fuel into a higher dust explosion class (Table 13).

The change to the next higher dust explosion class is imminent once the LEL of the flammable gas or vapor in air is reached; and is complete, once this limit is exceeded.

All dusts are categorized in the highest class once the concentration increases to, for example, the optimum concentration.

How do the explosion indices of cellulose change if, for example, butane or methane is added to the combustion air instead of propane? In order to make the corresponding comparison, the optimum concentrations for the explosion indices have been standardized as shown in Fig. 100. Again, it can be shown that the K_{St}-value is only marginally influenced with a linear dependency upon the flammable gas concentration, which in turn is partially influenced by the type of admixed gas. Propane and butane cause a much more pronounced increase in the explosion indices of cellulose than methane. The reason is that in a turbulent stage, propane and butane react more violently (p_{max} = 9.6 bar, K_G = 460 bar·m/s) than methane (p_{max} = 9.0 bar, K_G = 313 bar·m/s).

In summarizing, it can be stated that the optimum value for the maximum explosion pressure of hybrid mixtures consisting of combustible dust and flammable gases will occur slightly above the LEL. The maximum K_{St}-value of hybrid mixtures, however, occurs at the optimum concentration of the flammable gas addition.

The K_{St}-values coincide in general with the K_G-values of the flammable gas if brought to an explosion out of a turbulent state. But they can also be up to 15 % higher.

Table 13. Influence of propane content upon the dust classification into dust explosion classes

Propane content [vol%]		0	1.25[a]	2.5	4.25[b]
Dust category	K_{St} [bar · m/s]		Dust explosion class		
Pea meal	110	St 1	(St 1/St 2)	(St 2/St 3)	St 3
Cellulose	220	St 2	(St 2/St 3)	St 3	St 3
Optical brightener	295	(St 2/St 3)	St 3	St 3	St 3

[a] propane LEL with E = 10,000 J
[b] propane optimum concentration
( ) border line

4.3.3 Hybrid Mixtures

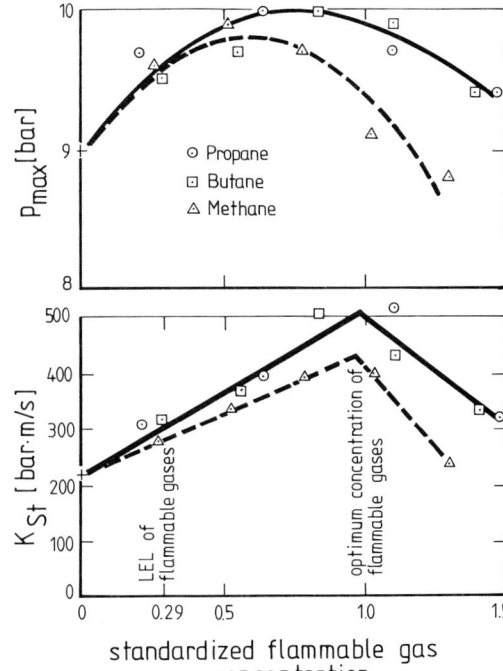

Fig. 100. Influence of the type of flammable gas addition upon the explosion indices of cellulose

Figure 101 shows the behavior of the optimum dust concentration for maximum explosion pressure and K_{St}-value at various flammable gas admixtures for hybrid mixtures with cellulose.

Again, the previously shown standard presentation was used for the flammable gas concentration. In general, the optimum concentration for the maximum explosion pressure decreases linearly with increased flammable gas concentration. For the K_{St}-value, this occurs only after the LEL of the added flammable gas has been markedly exceeded.

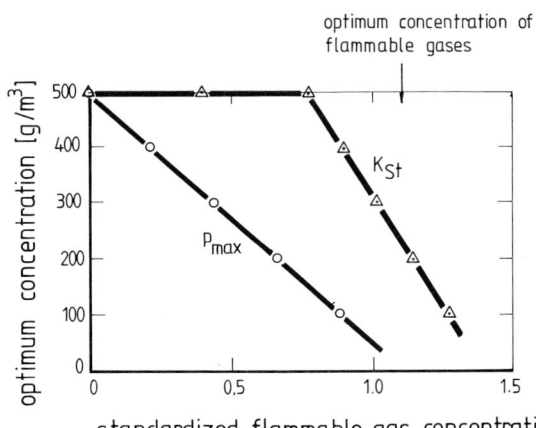

Fig. 101. Optimum dust concentration of hybrid mixtures (dust: cellulose)

4.3.4 Conclusions

Combustible dusts, as well as flammable gases, are only capable of exploding within a limited range, which is given by the lower explosible limit (LEL) and the upper explosible limit (UEL).

The LEL is of special importance for industrial practice because it determines whether dust explosions have to be anticipated once effective ignition sources are present. Then, constructive protective measures have to be selected. The LEL is only marginally affected by the applied ignition energy, but it is considerably dependent upon the process temperature. The maximum explosion pressure p_{max} and the maximum pressure rise dp/dt describe the violence with which dust explosions can occur. The maximum explosion pressure in closed, almost-spherical vessels of sufficient size (V≥20-l) is independent of the volume. The maximum pressure rise, however, is volume-dependent. It decreases with increased volume in accordance with the cubic law. The K_{St}-value which results from this law is specific for the dust and the test method but is independent of the vessel size for volumes V≥20-l. Both explosion indices are mainly influenced by the particle size and are only representative for the dust if the median particle size M ≤ 63 μm and the sample is tested in a dry state. Other parameters include product moisture, initial pressure, and temperature. The determination of the above-mentioned explosion indices has to be in accordance with a given test procedure because of the additional dependency upon the turbulence of the dust/air mixture at the time of ignition. Basically, such statements are also valid for flock as a combustible product. In such a case, the product of denier times length of cut, to measure the weight of the flock particle, will take the place of the usual median particle size M.

Special consideration has to be given to reactions of dust/air mixtures which occur in the presence of gas or vapor/air mixtures. With such hybrid mixtures, it has to be kept in mind that:

a) non-explosible dust and non-explosible gas or vapor/air mixtures can create explosible hybrid mixtures.
b) the maximum explosion pressure of combustible dusts will increase with admixture of flammable gases.
b) the K_{St}-value of combustible dusts will increase drastically and reach a K_{St}-value at optimum concentration which is equivalent to the one obtained when igniting under turbulent conditions. Therefore, a higher dust explosion class will result.

4.4 Safety Characteristics of Airborne Dust Describing the Ignition Behavior

4.4.1 Minimum Ignition Energy

4.4.1.1 Preliminary Remarks

The minimum ignition energy requirement for combustible dusts, flock, and hybrid mixtures is of great importance for the assessment of the danger in dust-producing plants. It determines the scope of the protective measure and its cost. This is especially true for the use of the protective measure "prevention of ignition sources" and also for the understanding of the ignition phenomena involving static electricity (spark discharges, brush discharges, propagating brush discharges, conical pile discharges) and mechanical sparks (short duration grinding and long duration friction sparks) [23].

The minimum ignition energy of a combustible dust, flock, or hybrid mixture in air is defined as the lowest electrical energy stored in a capacitor which, once discharged over a spark gap, barely ignites the most readily ignitable dust/air concentration at ambient temperature and atmospheric pressure. The discharge of the spark is extended through an inductance in the discharge circuit [66].

4.4.1.2 Apparatus for the Determination of the Minimum Ignition Energy

The apparatus shown in Fig. 102 can be used for the generation of sparks within an energy span of $E = 0.001-100$ J [7, 60, 63]. A peg board facilitates the systematic addition of capacitors, inductors, and resistors to the circuit which will determine the energy and the timely discharge of the spark.

The apparatus is designed for a maximum potential $E = 30$ kV and a maximum charging current $I = 1$ mA. The described ignition apparatus works in conjunction with three electrodes (Fig. 103) which form an auxiliary and a main spark gap with one common electrode. The auxiliary spark has a defined discharge time and a very low energy, which, for instance, is incapable of igniting the most readily ignitable propane/air mixture (minimum ignition energy = 0.27 mJ). The main spark is activated at a predetermined time. The arrangement of the capacitor discharge apparatus is shown schematically on the wiring diagram in Fig. 104.

Referring to Fig. 104, the selected capacitors are first charged to a high voltage $E = 8$ kV through a charging resistor $R = 1$ kΩ by closing the contacts of relay K_1. Relay K_1 opens immediately before the spark jumps so that the capacitors are separated from the high voltage source. Then the switch marked "Jennings" will be closed. However, the main spark cannot jump because its 6 mm gap is larger than the corresponding gap for the given voltage. The air has to be ionized in order to force the discharge across the gap. This is accomplished by the discharge of the auxiliary spark across the 2 mm gap, through interrupting the primary current of a high voltage inductance. The ionized gas molecules eliminate the insulation between the

Fig. 102. Capacitor discharge apparatus for the energy E = 0,001–100 J

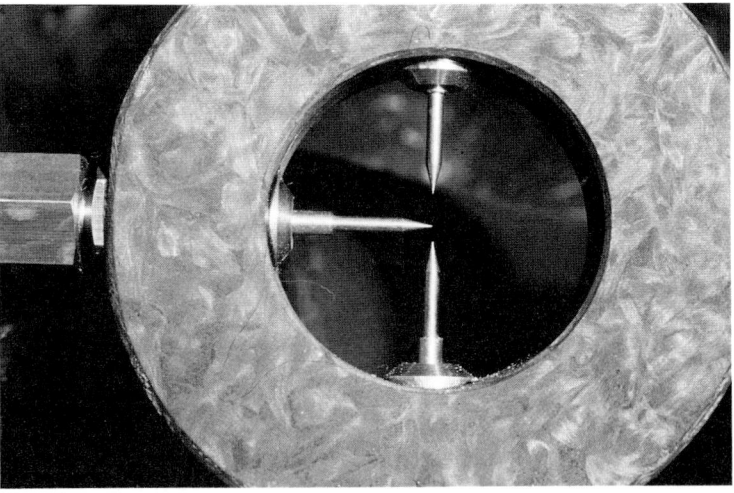

Fig. 103. Three-electrode arrangement for controlled capacitor discharges in laboratory equipment

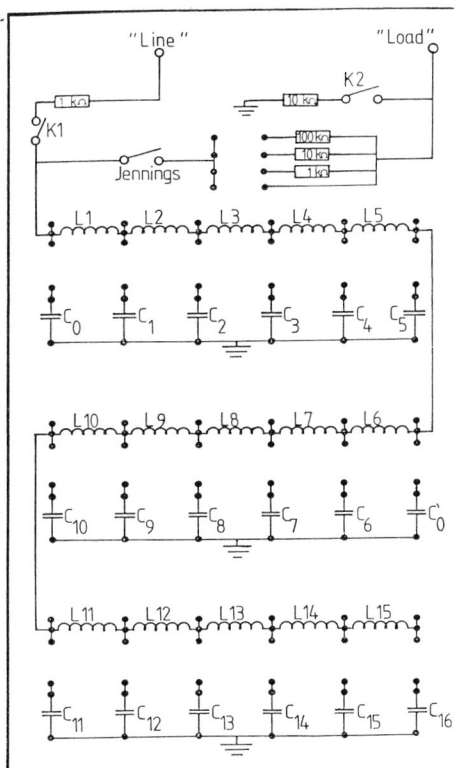

Fig. 104. Schematic wiring diagram for the capacitor discharge apparatus shown in Fig. 102

main electrodes, the ohmic resistance drops, and the charged capacitor C can discharge. The energy of the auxiliary spark is so small that the error in the energy statement for the main spark is negligible. The activation of the main spark is synchronous with the formation of the dust cloud which has to be ignited.

The potential energy E of the charged capacitor at the high voltage U and capacitance C is:

$$E = 1/2 \, C \, U^2$$

The charging and discharging process also charges additional capacitors which are provided by the cables, connectors for the electrodes, and the arrangement of the electrodes. They may affect the accuracy of the energy statement made above. In [60] it is documented that even for small stored energies (≤ 10 mJ) such an influence is negligible, and within the range of accuracy, provided the additional capacitance of the electrode arrangement is less than 200 pF.

4.4.1.3 Ignition Behavior of Combustible Dusts

In order to determine the minimum ignition energy of a combustible dust, a capacitor energy is selected which has a 100% probability for ignition. Then, at a constant dust concentration, the energy level is repeatedly cut in half until it reaches a value which will not result in a dust explosion for five consecutive ignition tests. Corresponding tests are carried out over a wide range of concentrations, which will reveal the concentration which is most readily ignitable. At such a concentration, the test is repeated until twenty consecutive tests will show no ignition.

If the most readily ignitable concentration cannot be defined, then the ignition tests have to be repeated over a wide range of concentrations on the basis of twenty non-ignitions. The actual minimum ignition energy of a combustible dust lies within an energy range given by the energy which just does not ignite the dust/air mixture and one which is 100% higher. With a justifiable laboratory effort, the minimum ignition energy for combustible dusts can only be stated as an energy interval (the exact value can only be determined with an extraordinary test effort). Therefore, the following distinctions will be made:

- The statement for the minimum ignition energy MIE refers to any dust concentration and is based on five non-ignitions.
- The statement for the lowest minimum ignition energy LMIE refers to the most readily ignitable dust concentration and is based on twenty non-ignitions. It will be identified as the lowest ignition energy of a dust.

The above definitions present a certain, but justifiable contradiction with the procedure [66] of recent date, as was already pointed out (see sec. 4.3.1.5). The ignition delay time t_d is the only measure for the turbulence of dust/air mixtures tested in a dispersed state. Short ignition delay times result in a high turbulence, long ignition delay times in a low turbulence.

Figure 105 shows the correlation of the minimum ignition energy with the dust concentration for one dust at various ignition delay times in the 20-l laboratory apparatus. Thus, one obtains a parabolic correlation between the minimum ignition energy and the concentration. The latter can only be stated (with a certain approximation) if the dust container is just emptied, i.e., at the ignition delay time used for the determination of the explosion indices $t_d = 60$ ms. For shorter ignition delay times ($t_d < 60$ ms), the referenced container is not yet emptied, for longer ignition delay times ($t_d > 60$ ms), however, the dust/air mixture may start to settle. At present, there is no technical possibility available to actually measure the dust concentration at the time of ignition. Therefore, the term "dust concentration" will be used in the following statements even if the ignition delay time t_d does not coincide with the time required to empty the dust container. As can be seen from Fig. 105, the most readily ignitable dust concentration increases with decreasing turbulence of the dust/air mixture. The lowest minimum ignition energy, however, decreases hyperbolically. This was confirmed through tests with five additional dusts (Fig. 106).

The lowest values for the lowest minimum ignition energy of combustible dusts occur at ignition delay times which are markedly longer than the test delay time for

4.4.1 Minimum Ignition Energy 97

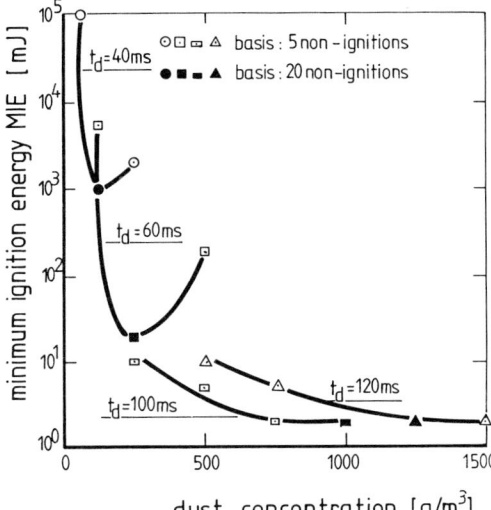

Fig. 105. Correlation of minimum ignition energy MIE with dust concentration (Lycopodium, 20-l laboratory apparatus)

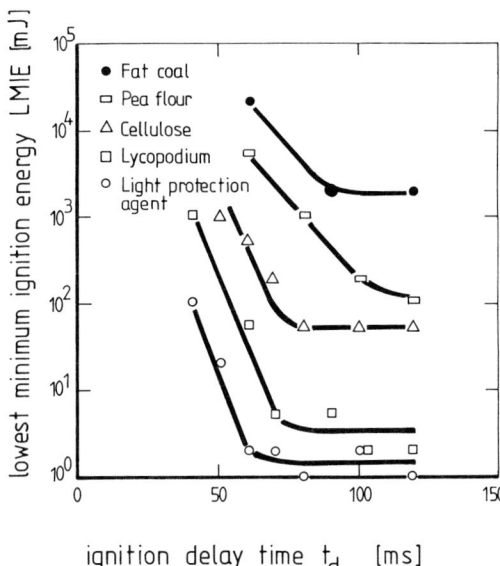

Fig. 106. Correlation of the lowest minimum ignition energy (LMIE) with ignition delay time t_d (20-l laboratory apparatus)

the explosion indices ($t_d = 60$ ms), that is, at a lower turbulence of the dust/air mixture. Therefore, an ignition delay time $t_d = 120$ ms is used in the laboratory apparatus for the determination of the energy boundary value.

In case a large vessel $V = 1$-m^3 is used for the determination of the LMIE, a similar correlation appears (Fig. 107), whereby the size of the dust container will

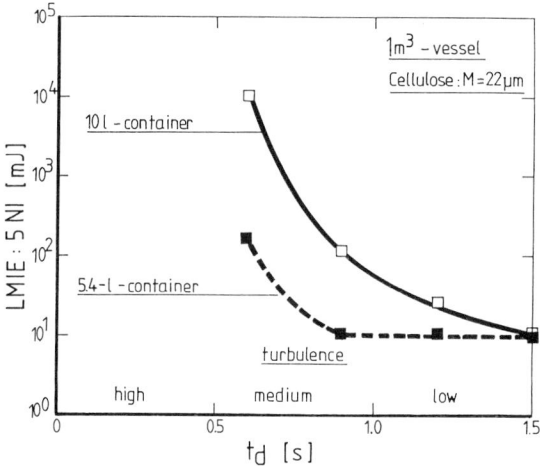

Fig. 107. Correlation of the lowest minimum ignition energy (LMIE) with ignition delay time t_d (1-m³ vessel)

have an additional influence (compare Fig. 62). For the 5.4-l container size, an ignition delay time $t_d = 1.2$ s is to be selected for the determination of the energy boundary value. For the 10-l size such a value is $t_d = 1.5$ s. Here again, the ignition delay times are distinct from the ones which have to be used for the determination of the explosion indices ($t_d = 0.6$ s versus $t_d = 0.9$ s)

In comparing the values of the minimum ignition energies gained in the 20-l laboratory apparatus and the 1-m³ vessel, using the prescribed test procedure, there is no difference within the range of accuracy (Fig. 8). Glarner [60, 63] has proven that the materials of construction for the electrodes (tungsten, stainless steel, brass, aluminium, lead) do not noticeably influence the ignition behavior of combustible dust; this, however, is not true for the distance of the main gap and the existence or non-existence of an inductance in the discharge circuit.

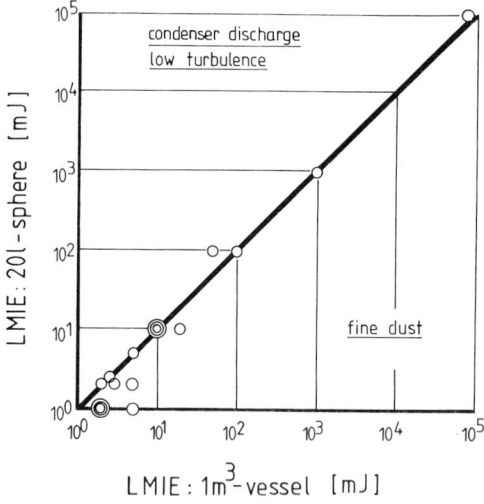

Fig. 108. Influence of the test vessel upon the lowest minimum ignition energy (LMIE)

4.4.1 Minimum Ignition Energy

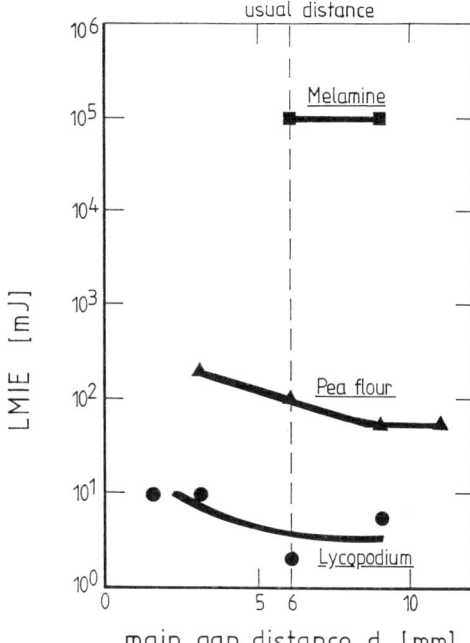

Fig. 109. Influence of the distance d of the main spark gap upon the lowest minimum ignition energy (LMIE)

With an increasing length of the main spark gap the value of the LMIE for combustible dust decreases (Fig. 109). Very short sparks which are generated with a discharge circuit without inductance (Fig. 110) are generally markedly less ignition-effective than the time-extended sparks which have an inductance in the discharge circuit. However, an inductance within the range of L = 1.32 – 46.4 mH does not exert a supplementary influence upon the ignition behavior of the dusts.

The LMIE measured with a plain capacitor discharge is ten times greater than the one obtained with the extended capacitor discharge in accordance with the test procedure [65]. Such a fact is of great importance in practice for the evaluation of the danger of, e.g., electrostatic ignition sources. The latter are pure spark discharges and their energies cannot simply be compared with the minimum ignition energy generated as per procedure. However, recent findings indicate that there are dusts, e.g., sulfur, which have a LMIE which is independent of the existence of an inductance in the discharge circuit. Therefore, the lowest results for the LMIE are only obtained with the following parameters for the electrode arrangement:

– pointed stainless-steel electrodes with a 2-mm diameter
– spark gap of the main electrodes: 6-mm
– inductance of the discharge circut L ≥ 1.32 mH
– no additional ohmic resistance in the discharge circuit.

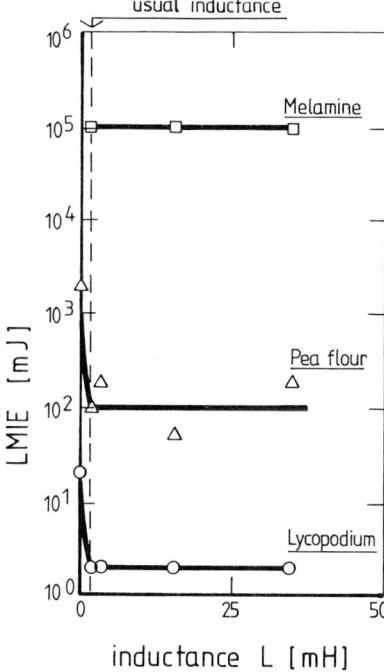

Fig. 110. Influence of the inductance L upon the lowest minimum ignition energy (LMIE)

Figures 111 and 112 show statistically the influence of the number of tests per energy upon the ignition probability for two differently ignitable, combustible dusts, at the most ignitable concentration. Both ranges are practically independent of the number of tests per energy provided it is ≥10.

That the 100% ignition probability is reached at a low ignition level with only five and not with more tests can be explained by the higher likelihood of a non-ignition in a longer test run.

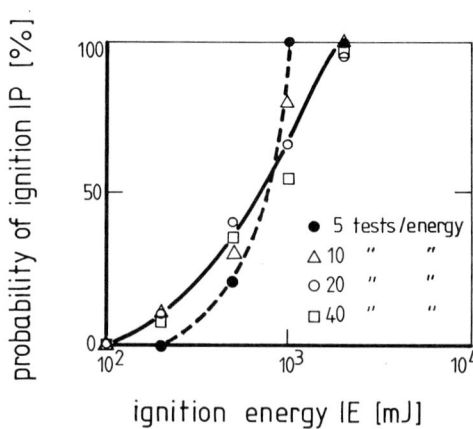

Fig. 111. Influence of the number of tests per ignition energy upon the probability of ignition (cellulose, readily ignitable concentration)

4.4.1 Minimum Ignition Energy

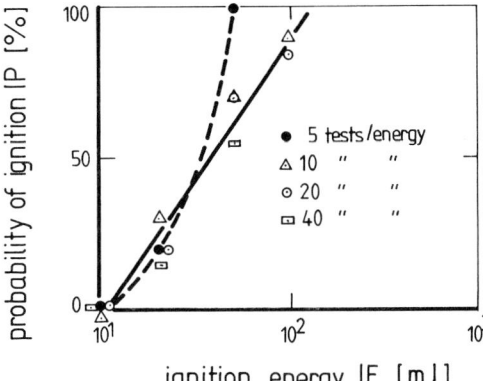

Fig. 112. Influence of the number of tests per ignition energy upon the probability of ignition (pea flour, readily ignitable concentration)

Furthermore, the determination of the LMIE for the easily ignitable cellulose (LMIE = 10 mJ) requires only five tests per energy level whereas the hard-to-ignite flour (LMIE = 100 mJ) requires at least 10 tests per energy level. An increase in the number of tests will in both cases not lower the energy level. As was already shown, the number of necessary tests increases with decreasing ease of ignition. It is, therefore realistic to test in accordance with the test procedure [66] for the LMIE with twenty tests per energy level at the readily ignitable concentration, especially for hard-to-ignite dusts.

Figure 113 indicates that the range of transition for the statistical probability is influenced by the ease of ignition of combustible dusts. The range expands with increasing energy and can be one thousand-fold for hard-to-ignite dusts (LMIE $\geq$ 100 J). This again explains why the stated high test energy (E = 10,000 J) has to be used for testing the explosibility of combustible dusts. Figure 114 shows the frequency of the LMIE obtained in testing approximately 200 dusts which is mostly in

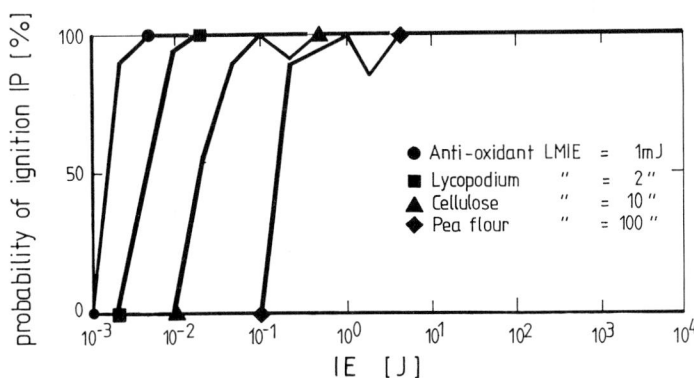

Fig. 113. Influence of the lowest minimum ignition energy upon the ignition probability

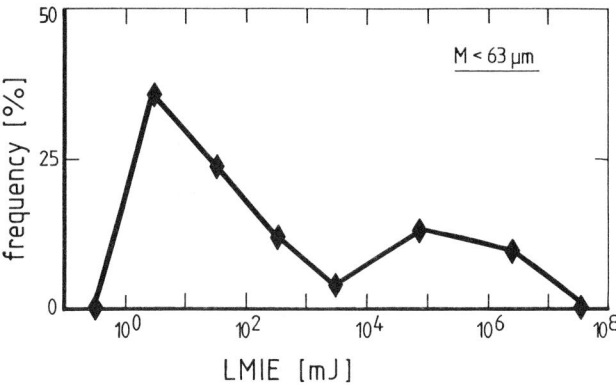

Fig. 114. Frequency of the lowest minimum ignition energy for approximately 200 tested dusts

the range of LMIE = 10 mJ. The frequency decreases with increased energy and then again increases. The latter seems to be true for mixtures of easy- and hard-to-ignite dusts and for mixtures of organic and inorganic dusts, based on current knowledge.

In Table 14, the explosion indices of some randomly selected fine dusts are compared with the lowest minimum ignition energy (LMIE).

Apparently there is no correlation between the maximum explosion pressure p_{max} or the K_{St}-value, on one hand, and the LMIE, on the other hand. There are dusts which, after the explosion is initiated, react weakly and which may have either a low energy boundary value (dicarbonic acid) or a relatively high energy boundary value (melamine polymer). This is also true for very violently reacting dusts. Bromphenoxime is very hard to ignite; hydroxyazobenzene, however, is very easy to explode with an extended capacitor discharge. Figure 114 and Table 14 show that the LMIE values of combustible dusts cover a wide range.

Table 14. Comparison of explosion indices with the lowest minimum ignition energy of fine dusts

Dust type	M [µm]	p_{max} [bar]	K_{St} [bar · m/s]	LMIE [mJ]
Dicarbonic acid	44	4.4	24	10
Melamine polymer	10	7.7	35	$100 \cdot 10^3$
Bromphenoxime	50	10.5	229	$250 \cdot 10^3$
Hydroxy-azo-benzene	<10	9.4	305	<1
Aluminum	<20	11.4	624	<1

4.4.1 Minimum Ignition Energy

Some dusts can only be made to explode with high energies (E ≥10,000 J) others with lower levels (E = 10 mJ) or very low energies (E <1 mJ). The latter are as readily ignitable as the flammable gases: methane, butane, propylene, or propane. Such a statement is contrary to earlier findings [67, 68]. According to these findings, the required energy to ignite dust/air mixtures should be one hundred to one thousand times higher than the ones needed to ignite normally ignitable flammable gases.

The prevention of ignition sources as the sole safety measure for equipment which processes and conveys especially easily ignitable dust should only be adopted with expert judgement in justifiable, exceptional cases [23].

The question can now be raised concerning which additional parameters influence the LMIE of combustible dusts [60]. Figure 115 shows the correlation between initial pressure p_i and the LMIE for an easily ignitable (Lycopodium), normally ignitable (pea flour), and hard-to-ignite dust (melamine polymer). In the range of 1–1.6 bar (abs) for the initial pressure, there is no noticeable influence. For negative pressures [p_i <1 bar (abs)], however, the readiness for dusts to ignite decreases. This is understandable because the most readily ignitable dust concentration decreases with decreased initial pressure which, in turn, may call for an increase in the energy boundary value. For the melamine polymer, such an influence was not observed for the pressure range tested, but one can assume that the LMIE will increase with an additional lowering of the initial pressure [p_i <0.6 bar (abs)].

If normal fluctuations occur around standard pressure in installations handling combustible dusts, a change in the ignition behavior need not be considered for the safety evaluation.

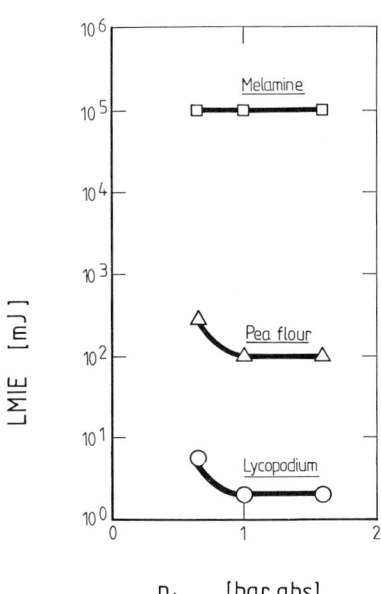

Fig. 115. Correlation of initial pressure p_i with lowest minimum ignition energy (LMIE)

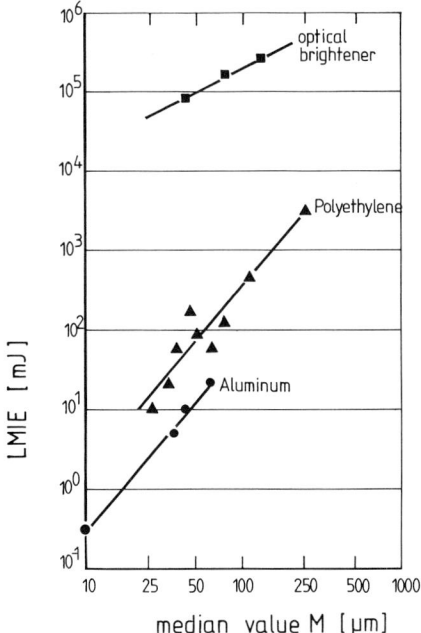

Fig. 116. Correlation of median particle size M and lowest minimum ignition energy (LMIE)

A critical inspection of the test results published in the literature [67] makes evident that often the median particle size or the particle distribution is not reported. Therefore, the impression is created that such a parameter has no noticeable influence upon the LMIE of combustible dusts. But this is not the case (Fig. 116): the finer the dust, the easier it can be ignited with an extended capacitor discharge. For polyethylene and aluminum the readiness for ignition changes with the third power of the median particle size. This is not the case for the optical brightener which had its energy boundary values determined with pyrotechnic ignitors of gradated energy levels. Such a type of ignition has a different effectiveness than the extended capacitor discharge. In order to characterize the ignition behavior of a combustible dust, it is therefore necessary to test its fine fraction.

Systematic tests pertaining to the ignition behavior of pure coating powders [69] with an organic content of 99% have shown that they have to be considered relatively easily ignitable if they are present as a fine dust, Fig. 117. Extruded, fine coating powders with an organic content down to 60% are similarly ignitable. The addition of aluminum decreases the LMIE (Fig. 118). Exploratory tests with epoxy powders have shown that the energy boundary value may be ten times larger (LMIE > 10 mJ) if a pure capacitor discharge is used.

Zeeuwen and van Laar [70] researched the influence of the water content of dusts upon the LMIE. They found that the effect of water is much more pronounced (Fig. 119) than that of the median particle size (Fig. 116). Therefore, the ignition behavior of combustible dusts has to be tested in the dry state.

4.4.1 Minimum Ignition Energy

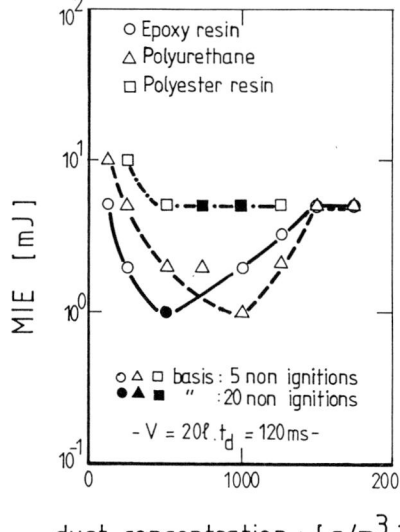

Fig. 117. Pure coating powders: Correlation of dust concentration with minimum ignition energy (MIE) (M <35 μm)

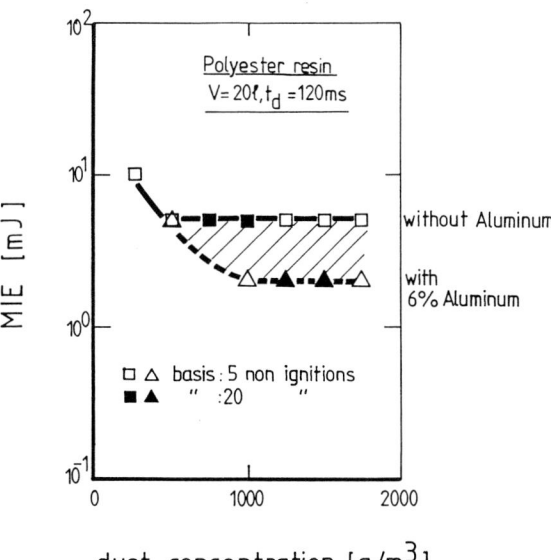

Fig. 118. Pure coating powders: Influence of added aluminum upon the correlation of dust concentration with minimum ignition energy (MIE) (M = 34 μm)

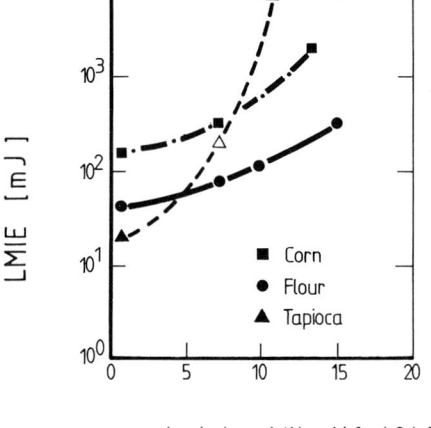

product humidity H [wt%]

Fig. 119. Influence of the humidity (water content) of combustible dusts upon the lowest minimum ignition energy (LMIE) [70]

Another parameter which has a major influence upon the ignition behavior of combustible dusts is the temperature. Glarner [60] observed that the LMIE for hard-to-ignite dusts (melamine) decreases much more rapidly with increased temperature (Fig. 120) than for easily ignitable dusts (Lycopodium). He noticed that the straight lines in a double logarithmic plot intersect at one point having the coordinates [1000°C; 0.088 mJ]. The boundary energy level referenced to a temperature T = 1000°C coincides with LMIE = 0.088 mJ for all tested dusts. From today's view point, the mechanism of a dust explosion can roughly be explained as the combustion of the volatile components at the particle surface (see sec. 2.2.). Glarner's interpretation therefore suggests that the smoldering gases driven-off have a similar ignition behavior towards an extended capacitor discharge. This is parallel to the ignition behavior of the homologous flammable gases of the alkanes, which also have a similar lowest minimum ignition energy (LMIE = 0.24–0.31 mJ).

It also must be mentioned that the assumption for the calculation of the LEL of combustible dusts [53] is as follows: a complete flame propagation through the entire explosion volume is only possible if the energy liberated through the reaction is capable of heating the products of the reaction, the excess air and all other ballast to approximately 1000°C. This again is an interesting parallel to the common intersection of the minimum ignition energy at a temperature of 1000°C.

A substantially erroneous estimate is possible if the MIE values determined at room temperature are used for the assessment of the dangers of electrostatic ignition sources for installations which process and handle combustible dusts at elevated temperatures. On one hand, the importance of the MIE of combustible dust for safety evaluations is undisputed; on the other hand, its determination requires a large test effort which will increase further if temperatures higher than room temperature are studied. Therefore, it became necessary to study whether a statement about the range of the energy boundary was possible from, e. g., the determination of the normal explosion indices. Such a path was indeed found [7, 60, 63]. Every

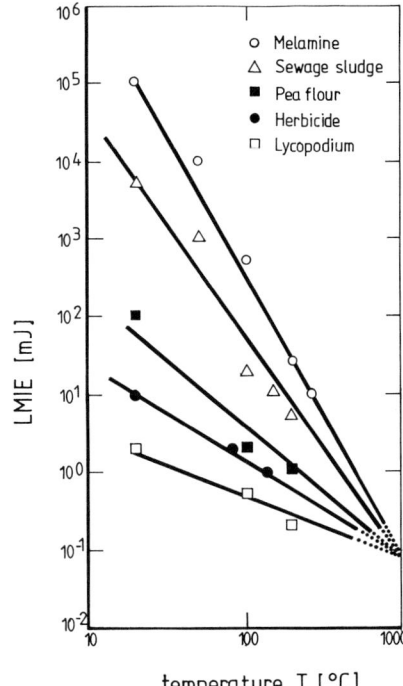

Fig. 120. Influence of the temperature upon the lowest minimum ignition energy (LMIE) [60]

test for the determination of the explosion indices of a combustible dust which is done with two pyrotechnic ignitors having a total energy of E = 10,000 J renders automatically the time intervals for the duration of combustion t_1 and induction t_2. The definitions of these parameters are shown in Fig. 121 and are as follows:

t_1 – the duration of combustion – is the time interval between ignition and the attainment of the maximum explosion pressure.

t_2 – induction time – is the time interval between ignition and the intersection of the tangent at the point of inflection of the pressure curve with the abscissa through the initial pressure.

The time indices which are defined above are principally valid for any dust concentration. The following statements refer to the minimum values which are obtained during the normal testing for the dust explosion indices.

For more than fifty dusts which have been tested in the 1-m³ vessel and the 20-l laboratory apparatus, the following unequivocal correlation has been shown:

The LMIE and the shortest induction time or the minimum combustion time can be correlated (Fig. 122) when plotted on a double logarithmic scale. The times have to be derived from tests using the usual IE of 10,000 J.

A dust can be classified as very easily ignitable (LMIE <10 mJ) if the minimum times indicated in Fig. 122 are less than 50 ms for the normal test results gained in

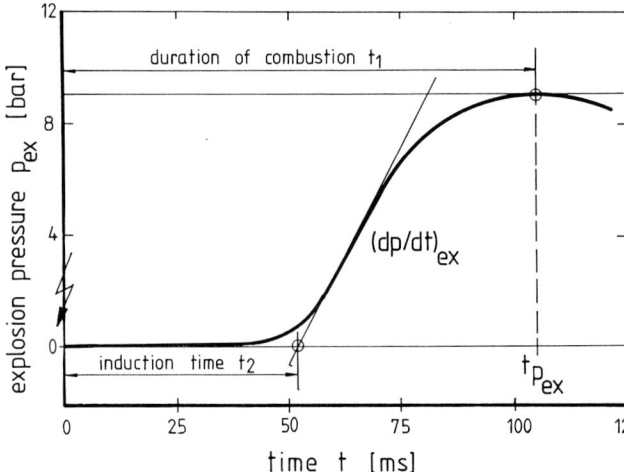

Fig. 121. Definition of the duration of combustion t_1 and induction time t_2

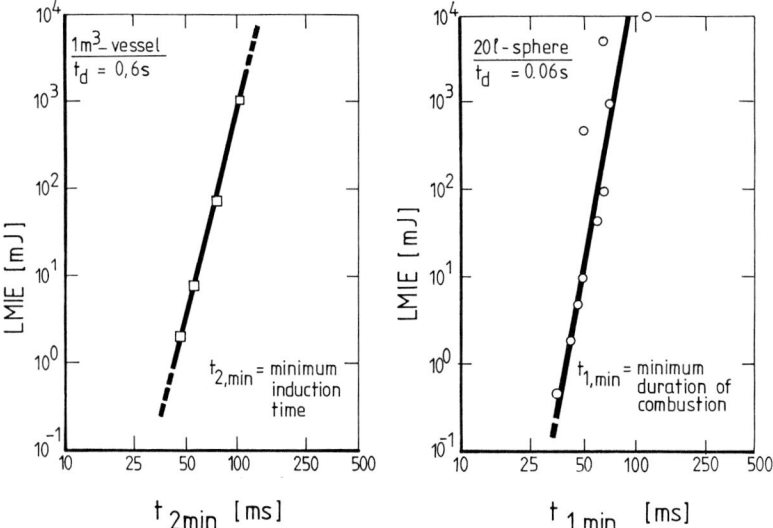

Fig. 122. Correlation of the lowest minimum ignition energy with the time indices $t_{1,min}$ and $t_{2,min}$ for combustible dusts

the 20-l or 1-m³ test apparatus using the given test procedure. A similar correlation exists for the LMIE [60, 63] at a longer ignition delay time (t_d = 120 ms).

The correlation shown in Fig. 122 is also valid at varying initial pressures, at differing oxygen/nitrogen concentrations in the air of combustion and at changing temperatures.

4.4.1 Minimum Ignition Energy

In order to circumvent the determination of the time indices at elevated temperatures, it is easier to reproduce the temperature dependence of the LMIE of the dust in accordance with Fig. 120 by drawing a straight line through the coordinates [25 °C; LMIE] and [1000 °C; 0.088 mJ]. This is a fast and easy method for the determination of the range of the LMIE of dusts. But it is only valid if the explosion indices are independent of the applied ignition energy within the range of accuracy (Fig. 85). If this is not the case, then the energy boundary value can be estimated from the linear behavior of the K_{St}-value with the ignition energy (Fig. 86).

4.4.1.4 Ignition Behavior of Flock

For the determination of the LMIE of flock, the arrangement of the electrodes has to meet the same conditions as described in conjunction with combustible dusts (see sect. 4.4.1.3).

Extensive systematic tests for the LMIE for flock [62, 63, 71] showed, analogous to dusts, that a parabolic correlation exists for the MIE and the flock concentration, which is additionally influenced by turbulence of the mixture. Low turbulence, i.e., longer ignition delay times, result in the lowest value for the boundary energy (Fig. 123).

Similiar to the explosion indices, the LMIE of flock is mainly influenced by the product denier times cut length (Fig. 124). The type of flock (cut or ground flock) or the type of fiber are of minor importance. Table 15 shows the pertinent correlations.

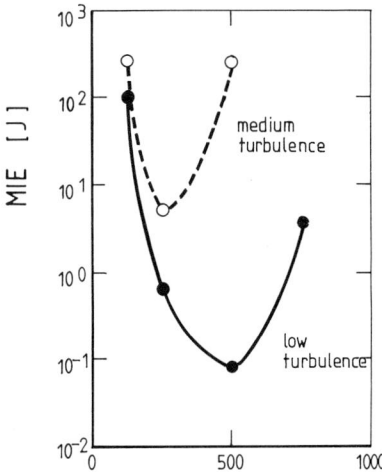

Fig. 123. Correlation of flock concentration with minimum ignition energy (MIE) (3.0 D/0.5-mm)

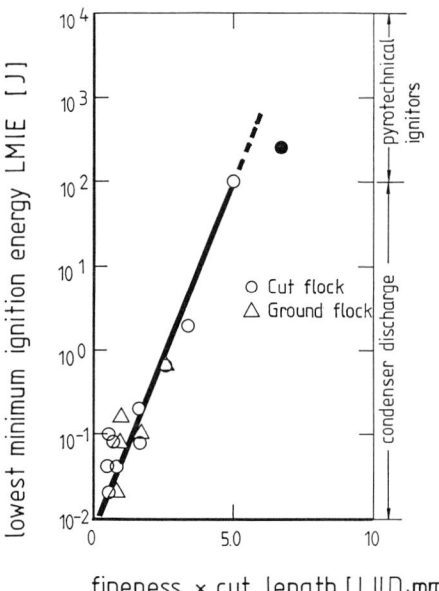

Fig. 124. Correlation of lowest minimum ignition energy (LMIE) of flock with the product denier times length of cut

Table 15. Correlation of lowest minimum ignition energy LMIE of flock with the product denier times cut length

1.11 denier × cut length in mm	LMIE [J]
0.49	0.020
1.53	0.200
2.97	4.000
6.03	> 100

Very extensive testing with flock allows a statement with regard to the influence of ignition energy upon the statistical transition of the ignition probability (Fig. 125). Analogous to combustible dusts (Fig. 114), the range of transition increases with increased LMIE. For a flock with a high value of the product denier times cut length a very potent ignition source is required in order to recognize instantly its explosibility. In performing the normal explosion tests in, e.g., the 1-m³ vessel, the time indices are obtained automatically (see Fig. 121 for the definition). Therefore, a linear correlation exists on a double logarithmic plot for the LMIE versus the minimum induction time $t_{2,min}$. This allows an estimate of the energy boundary value in an easy fashion, similar to combustible dusts (Fig. 126).

4.4.1 Minimum Ignition Energy

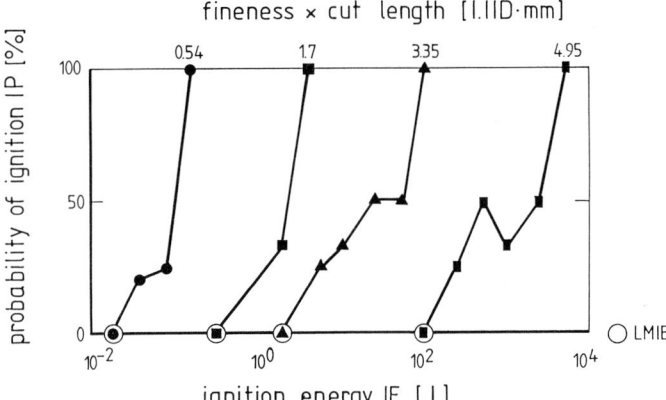

Fig. 125. Influence of the lowest minimum ignition energy upon the statistical transition of ignition probability

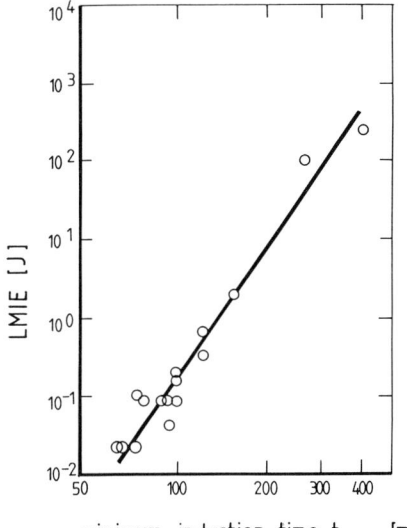

Fig. 126. Correlation of lowest minimum ignition energy (LMIE) of flock with minimum induction time $t_{2,min}$ (1-m³ vessel, low turbulence)

Flock machines may generate pure capacitor discharges generally not exceeding 500 mJ. Therefore, it was of interest to know the LMIE of flock without inductance in the discharge circuit. The results of such tests are shown in Fig. 127.

The LMIE of a fine easily ignitable flock increases approximately one hundredfold, which is slightly higher than anticipated from the dust results (Fig. 110). If such a result is confirmed through additional tests, then no concentration limitation will be needed (see sect. 4.3.2.2) in flock machines which have the above-mentioned limited energy for spark discharges.

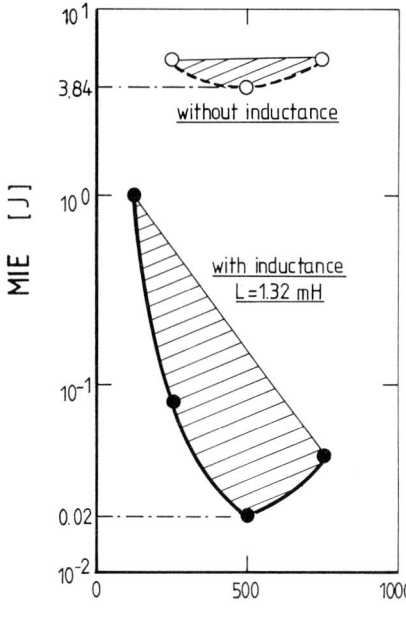

Fig. 127. Influence of inductance upon the lowest minimum ignition energy of acrylic flock (0.54 D/0.9-mm)

4.4.1.5 Ignition Behavior of Hybrid Mixtures

Earlier tests made by Franke [72] with methane and coal dust show that even methane concentrations below the LEL markedly reduce the LMIE of coal dust. Everywhere in industry where hybrid mixtures of combustible dust and flammable gases or vapors exist, the question will often be raised asking how much easier such mixtures can be ignited than pure dust/air mixtures. The answer to such questions is given by the experimental investigations made by Pellmont [7]. These are of utmost importance for the sole use of the safety measure prevention of effective ignition sources in practice.

Figure 128 shows the influence that an additional propane content in air has on the LMIE of hybrid mixtures with five dusts having different ignition potential. It shows that the energy boundry value of the dust also determines the ignition behavior of the hybrid mixture. On a semi-logarithmic plot there is a linear correlation with the additional gas content which is important. All lines intersect at the point which corresponds to the LMIE of the flammable gas. With a known LMIE of the combustible dust, additional flammable gas content and its most readily ignitable concentration, the ignition behavior of the hybrid mixture can easily be described. At the same time, the most readily ignitable dust concentration of the hybrid mixture decreases linearly with increased flammable gas content [7]. Therefore, the

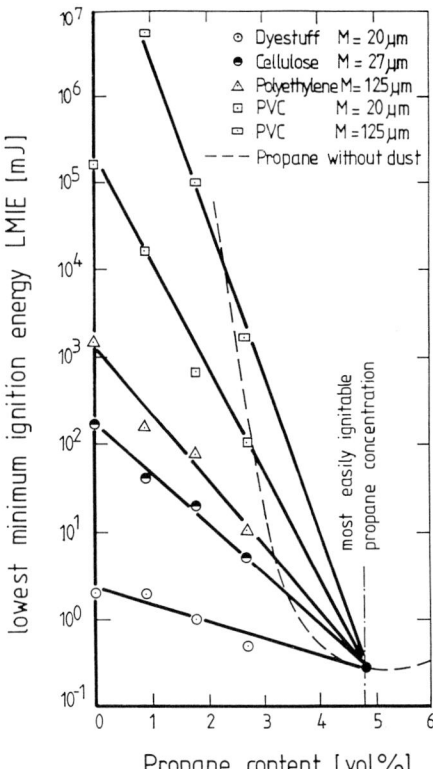

Fig. 128. Lowest minimum ignition energy (LMIE) of hybrid mixtures of combustible dust and propane

conclusion is reached that not so potent sparks (E < LMIE) which normally are incapable of igniting pure dust/air mixtures will ignite hybrid mixtures.

But not only is the ignition capability of dust increased with the addition of flammable gas, conversely the ease of ignition of certain propane/air mixtures can also be enhanced by the admixture of easily ignitable dust, as shown in Fig. 128. An energy of E = 100 J is required to ignite a propane/air mixture slightly above the LEL. However, in the presence of a dyestuff with a very low ignition energy, this minimum energy level decreases to E = 10 mJ. The question pertaining to the fignition behavior of hybrid mixtures in the presence of propane or similarly ignitable flammable gases or vapors can be answered as follows:

- In processing normal or hard-to-ignite dusts (LMIE ≥ 100 mJ), the flammable gas content may reach the LEL without resulting in a lowest minimum ignition level for the hybrid mixture, which is equivalent to the energy of a brush discharge.
- In processing easily ignitable dusts (LMIE ≤ 10 mJ), the minute addition of flammable gas (solvent vapor) will intensify the danger potential. Although not experimentally proven, the ignition of such hybrid mixtures by brush discharge is conceivable.

- The two ranges described above for the minimum ignition energy of combustible dusts cannot be unequivocally separated from the ignition behavior of their hybrid mixtures. Additional safety considerations are required relative to the ignition capability of brush discharges.
- If the content of flammable gas or vapor, however, exceeds the LEL, then an easy ignition which is within the ignition capability of brush discharges can basically be expected.

4.4.1.6 Conclusions

The lowest minimum ignition energy (LMIE) and the minimum ignition energy (MIE), respectively, represent a quantity which allows a comparison of the ignition behavior of combustible dusts towards an extended capacitor discharge, as defined [65]. Contrary to earlier assumptions, newer investigations have shown that a substantial number of dusts are relatively easy to ignite. Some very easily ignitable dusts are as ignitable as normally ignitable flammable gases, e. g., methane, propylene, propane (Fig. 129).

The LMIE of combustible dusts is, within the range of standard pressure, practically independent of the initial pressure (starting pressure of the explosion). But it is very dependent upon the median particle size, the temperature of the mixture,

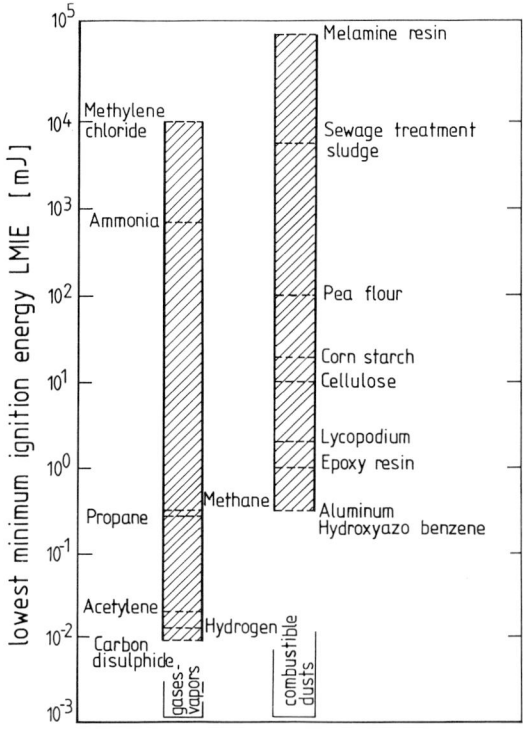

Fig. 129. Comparison of the lowest minimum ignition energies: combustible dusts, flammable gases and vapors

and the moisture content of the product. The range of the boundary energy can be estimated with sufficient accuracy from the minimum combustion time and minimum induction time, which routinely result from the normal dust explosion tests.

The values for the LMIE are based on extended capacitor discharges. These are generally much more ignition-efficient than pure capacitor discharges. A conversion to practical conditions is only possible if there the capacitors also discharge with an inductance present. But this is normally not the case. The capability of brush discharges to ignite flammable gas/air mixtures has been confirmed [73]. Tests have shown that they have the same ignition effectiveness as extended capacitor discharges with a maximum energy of 5 mJ. However, up to now it has been impossible to ignite easily ignitable dusts (LMIE <10 mJ) by a brush discharge which supposedly is due to the lack of induction. Therefore, it can be expected that brush discharges will only ignite dust/air mixtures which have a LMIE substantially below 1 mJ. The investigations are not yet concluded.

The above statements are basically also valid for the corresponding boundary energy of flock as a fuel. In such a case, the product denier times cut length has to be taken as the reference instead of the median particle size M which is customary for dusts. Here again, the values for the LMIE cannot be transferred to industrial practice because only pure capacitor discharges occur in flock machines. An estimate of the range of ignition energy is also possible from the times which are generated in the standard explosion tests.

The LMIE of dusts in hybrid mixtures is lowered differently, depending upon the admixture of flammable gases or vapors. This is deemed most dangerous for already easily ignitable dusts. The ignition behavior of hybrid mixtures can be described as a straight line in a semi-logarithmic plot which connects the LMIE of the dust with the one for the flammable gas at its most easily ignitable concentration.

4.4.2 Ignition Temperature

4.4.2.1 Preliminary Remarks

The characteristic value of the ignition temperature describes the ignition behavior of an airborne dust on hot surfaces. It is defined as the lowest temperature of such a surface which will just ignite the most readily ignitable dust mixture in air [23].

Hot surfaces represent 6% of the effective ignition sources responsible for actual dust explosions [74]. It has been shown that the ignition temperature and the LMIE determine whether the fuel-air mixture can be ignited with mechanically generated sparks. (They represent 28% of the ignition sources for actual dust explosions [74]). This documents the importance that the safety characteristics have for the assessment of ignition potential in industrial installations.

4.4.2.2 Apparatus for Temperature Determination

The German Federal Institute for Material Testing [Bundesanstalt für Materialprüfung, BAM] in Berlin [27] developed an apparatus for the determination of the ignition temperature of airborne dust (Fig. 130). It is known by the name BAM-oven.

This apparatus consists basically of a horizontally supported pipe-oven. Approximately 1–2 ml of the dust to be tested is blown against a heated parabolic plate by means of a manually operated aspirator bulb. The temperature of the heating surface and deflecting surface are monitored by thermocouples. The accuracy is $\pm 5\%$. The ignition tests are started after the oven has been preheated to 600°C. Then the heater is shut off and, at dropping temperature, the given amount of dust injected at 50°C intervals.

If there is no ignition at a certain temperature, then the oven is heated to the next higher level and the process repeated at falling oven temperatures and 10°C intervals. An ignition is recorded if the injected dust ignites with a bang or flame appearance.

The lowest ignition temperatures are only obtained if the described ignition test is carried out with systematically gradated dust amounts.

For completeness sake, it has to be mentioned that a so-called Godbert-Greenwald oven is used in the USA and Great Britain for the determination of the ignition temperature of combustible dusts. The dust sample is injected from the top into the vertically standing oven. The temperatures measured with such apparatus are in general substantially higher than the ones from the BAM oven [27].

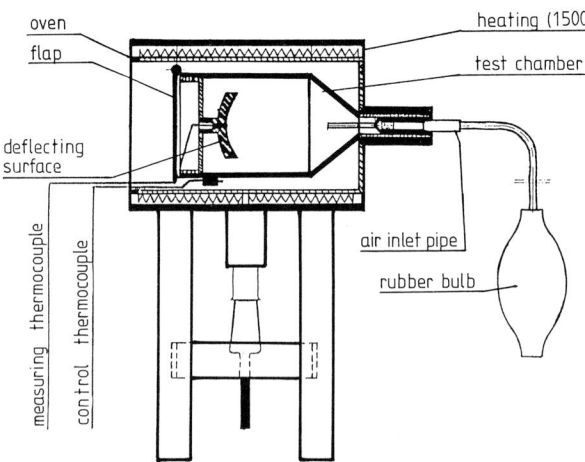

Fig. 130. Apparatus for the determination of the ignition temperature of combustible dusts (BAM-oven: schematic)

4.4.2 Ignition Temperature 117

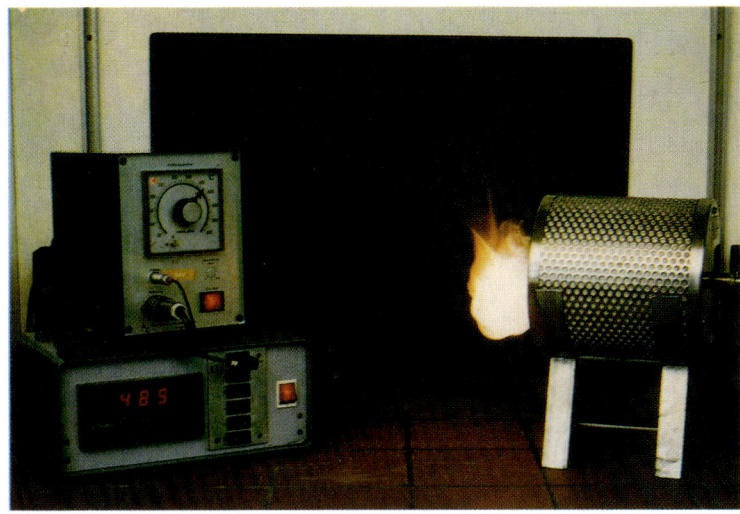

Fig. 131. Ignition of a dust/air mixture in the BAM-oven

4.4.2.3 Ignition Effectiveness of a Glowing Coil

For the qualitative screening of the explosibility of dusts in the modified Hartmann apparatus (Fig. 21), a glowing coil is used with a temperature $T_{Gl} = 1200\,°C$ (Fig. 132). This intensifies the ignition conditions in comparison to the continuous induction spark usually used [28].

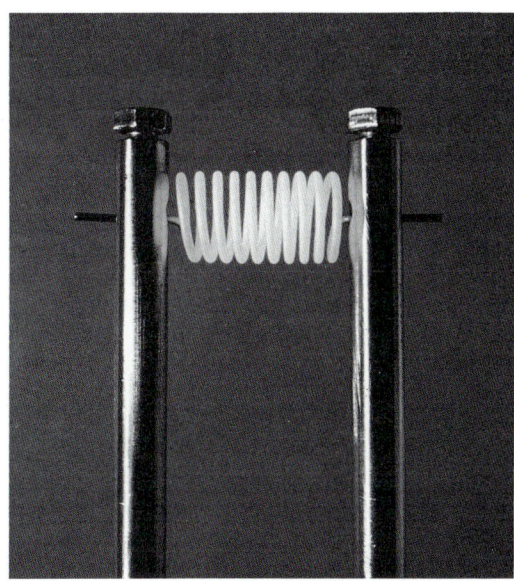

Fig. 132. Glowing coil as the ignition source for the modified Hartmann apparatus ($T_{Gl} = 1200\,°C$)

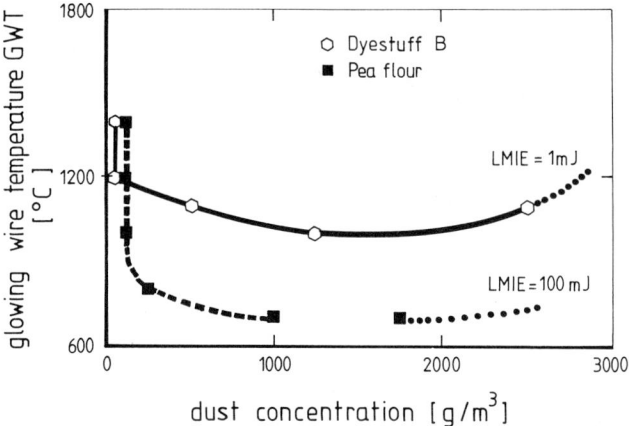

Fig. 133. Influence of the glowing coil temperature upon the flammable range of combustible dusts with varying lowest minimum ignition energy

The coil is made out of Kanthal A-1 (22 Cr, 5.5 Al) with a wire diameter of 1.2 mm which corresponds to an area of 1.13-mm^2. The straightened wire measures 470 mm and has a resistance of 0.6 Ω at 20 °C. The coil has ten turns with a diameter of 13.5 mm and is electrically heated. Glarner [60] has documented with systematic tests that there is no correlation of coil temperature with the LMIE of dusts (Fig. 133). For example, dyestuff B (LMIE = 1 mJ) which is very easily ignitable with an electric spark, requires a coil temperature of around 1000 °C for the ignition of its mixtures. Pea flour (LMIE =100 mJ), however, which is much harder to ignite, can be made to explode over a wide range of concentrations with a coil temperature T_{Gl} = 700 °C.

Glarner, however, found a unequivocal linear correlation between the glowing coil temperature which was necessary to ignite dust/air mixtures and the ignition temperature in the BAM-oven. He states absolutely that at the prescribed glowing coil temperature of T_{Gl} = 1200 °C, only such dusts will be recognized as explosible which have an ignition temperature T_{BAM} ≤ 540 °C. As some combustible dusts may also have a higher ignition temperature, a misjudgement of the dust explosibility is quite possible when the glowing coil is used as a test tool.

4.4.2.4 Conclusions

The ignition temperature of a combustible dust is an important characteristic. It not only describes the ignition behavior of mixtures on hot surfaces but also indicates, together with the LMIE, whether mechanically generated sparks will be capable of igniting dust/air mixtures. More on this issue will be discussed in a later paragraph.

In addition, it has been documented, that no correlation exists between ignition behavior of dust/air mixtures with extended capacitor discharges and the temperature of the glowing coil which is required for the ignition of the mixtures. The effectiveness of the two sources cannot be compared as the ignition source is in one case an electrical discharge and in the other a "hot surface".

4.5 Safety Characteristics of Airborne Dusts Describing the Course of an Explosion in Pipelines

In order to better understand the course of an explosion of combustible dusts in pipelines the behavior of flammable gases is first examined. For test purposes, such mixtures are generally ignited in a quiescent state. A distinction has to be made between two limiting cases [75, 76, 77].

1. Ignition of the explosive mixture at the open end of the otherwise closed pipe.

In this case, the flame travels first into the pipe which is filled with the gas mixture and the whole mixture burns in the pipe. The normal combustion velocity will not exceed a few cm/s or m/s, depending upon the type of flammable gas, because the products of combustion can flow away. With continuing combustion, however, the gas mixture contained in the pipe will be excited to its natural frequency, and the gas motion will not be laminar but turbulent. Therefore, the normal combustion velocity will increase. The flame front is no longer planar but concave, and the combustion area larger than the pipe section. With such test conditions, the normal combustion velocity is slightly increased by the quotient combustion area over pipe section.

2. Ignition of the explosive mixture at the closed end of the otherwise open pipe.

In this case the combustion velocity is substantially larger than with the ignition at the open end. Due to the major volume increase while burning, a so-called "displacement velocity" will affect the unburnt mixture ahead of the flame front, which will not only increase the turbulence but also the combustion area. Only a small portion of the gas mixture burns in the pipe. The rest is pushed through the open end and gets ignited outside the pipe by the trailing flame.

A third case exists for the combustion of gas/air mixtures with ignition at the closed end of a completely closed pipe.

During the first phase of combustion, i.e., initially after ignition, the normal combustion velocity is again influenced by the already-mentioned displacement velocity with the consequent increase in turbulence relative to the ratio combustion surface/pipe section. This combustion velocity is therefore relatively high. In the last phase of combustion, the displacement velocity, and, therefore, the turbulence, both decrease towards the closed pipe end, finally approaching zero, which means that the combustion velocity is also decreased. But this is only the case for relatively slow combustion reactions of, for instance, methane or propane. For the fast-burning hydrogen, such a decrease in the combustion velocity towards the pipe end is not noticeable.

What has been said for flammable gases is also applicable for the course of an explosion of dust/air mixtures in pipelines. Here, however, some dust dispersion and, therefore, some turbulence must exist in order to produce the mixture. The effects of the described displacement together with resulting turbulence will be superimposed. As indicated earlier (Figs. 59 and 60), the maximum explosion pressure p_{max} and the maximum pressure rise $(dp/dt)_{max}$ describe unequivocally the course of a dust explosion in closed cubical vessels. For pipelines, however, it has to be noted that the entire course of combustion has superimposed oscillations.

The determination of the latter dust explosion characteristics from the time-pressure oscillogram is quite difficult. Therefore, the flame velocity v_{ex} based on a certain pipe length (meter marker), is used to describe either the course of an explosion in pipelines or the maximum flame velocity v_{max}.

If the ignition of a dust/air mixture is initiated at the open end of a closed pipe, then the explosion will propagate with a low explosion velocity and weak pressure build-up towards the closed end, independently of the pipe diameter. If, however, the ignition point is relocated at the closed end of the otherwise open pipe, then the

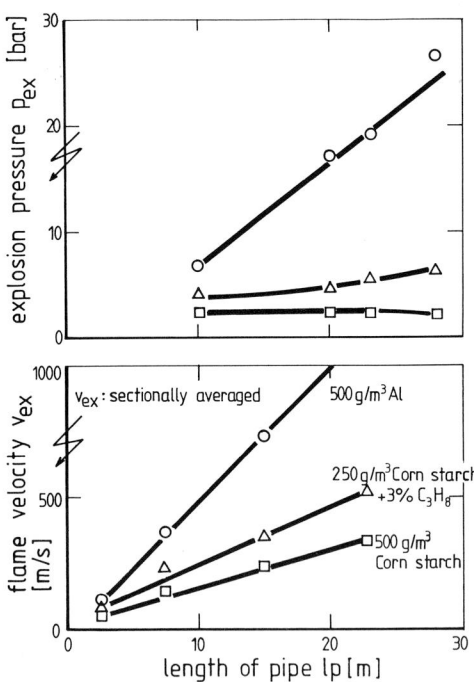

Fig. 134. Explosion characteristics of combustible dusts in a pipeline closed at one end (DN = 400-mm, lp = 30-m; ignition: closed end) corn starch: p_{max} = 9.7 bar, K_{St} = 210 bar.m/s, corn starch/propane: p_{max} = 9.5 bar, K_{St} = 412 bar.m/s, aluminum: p_{max} = 11.4 bar, K_{St} = 625 bar.m/s

flame velocity v_{ex} increases with increased length travelled. Figure 134 depicts such a behavior for two dusts and the hybrid mixture corn starch and propane in a pipeline having a 400-mm diameter. The velocity increase becomes larger as the dust-specific characteristic K_{St} of the fuel increases. This trend is also true for the explosion pressure p_{ex} (except for corn starch).

If the testing is carried out over a wide range of concentrations, then there is a linear correlation between the maximum flame velocity v_{max} and the maximum explosion pressure p_{max} in the pipeline, as shown in Fig. 135.

If the dust explosion occurs in a pipeline with a large diameter, the secondary explosion of the displaced unburnt mixture will be especially violent at the given ignition location (Fig. 136).

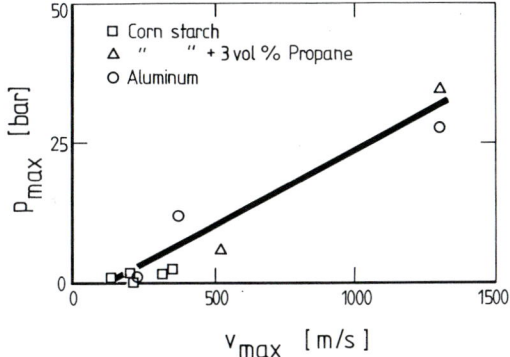

Fig. 135. Optimum explosion characteristics of combustible dusts in a pipeline closed at one end (DN = 400-mm, lp = 30-m, ignition: closed end)

Fig. 136. Flame propagation of a dust explosion in the vicinity of the opening of a pipeline (DN = 1600-mm, lp = 10-m, ignition: closed end)

122   4 Material Safety Specifications

For dust explosions in pipelines closed at both ends, the displacement velocity and turbulence have the same influence as described for flammable gases (Fig. 137). Corn starch explosions will not even reach the pipe end opposite the ignition source. Hybrid reactions, however, result in a constant terminal flame velocity. The flame velocity of the very violently reacting aluminum dust increases with increased distance, although the maximum value reached is markedly lower than the one for the pipe closed only at one end (see Fig. 134).

At first, the explosion pressure p_{ex} is independent of the point of measurement and is only influenced by the dust-specific characteristic K_{St}. Very high pressure values are registered in the area of the end flange opposite the ignition source. This is caused by the explosion flame running into the pre-compressed dust/air mixture. Pressures and dust-specific characteristics correlate linearly. The reaction of dust/air mixtures in pipelines is especially violent if they are ignited by a flame jet which has originated from a dust explosion in an upstream explosion vessel. In such a case, the mixtures are not only dispersed but also in motion.

Figure 138 may serve as a guideline. It shows the correlation of the maximum explosion pressure p_{max} and the maximum flame velocity v_{max} with the dust-specific characteristic K_{St} for flame jet ignition in a pipe having a nominal diameter of 400 mm and varying lengths.

It may be noted that the pipe length has a decisive influence upon the explosion. For lengths in the range of 20 to 40-m, explosions of combustible dusts with the dust-specific characteristic $K_{St} > 200$ bar.m/s (wood flour, pigment, aluminum

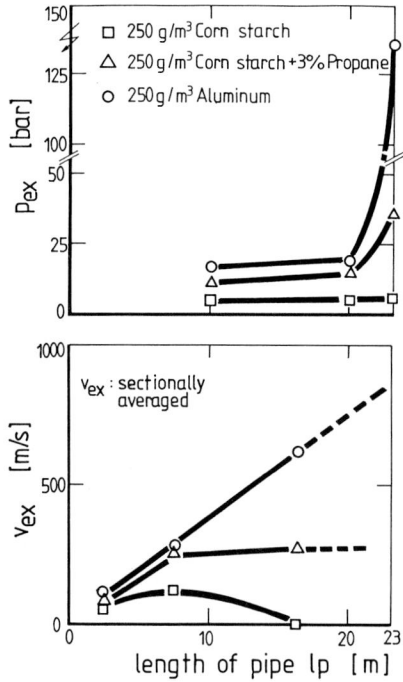

Fig. 137. Explosion characteristics of combustible dusts in a pipe line closed at both ends. (DN = 400-mm, lp = 23-m; Ignition at one pipe end, for characteristics see Fig. 134)

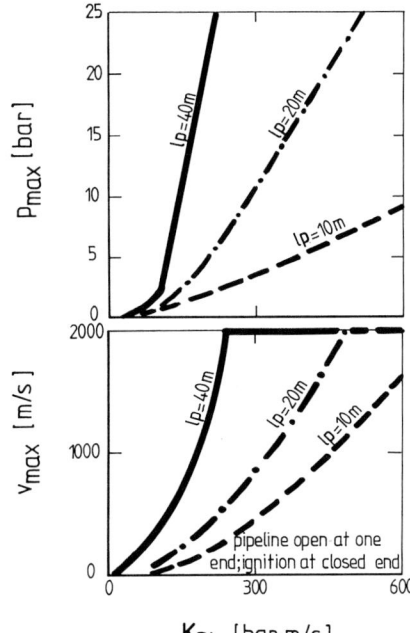

Fig. 138. Explosion characteristics of combustible dusts in a pipeline. DN 400-mm with flame jet ignition

dust) may convert into detonations with velocities in the range of 2,000 m/s. The resulting pressures are equivalent to the ones from gas detonations. A detonation of combustible dust is defined as a combustion process which maintains a constant high velocity over a long length. At the moment it cannot be decided whether it involves a stable detonation over a long pipe length or an unstable quasi-detonation. As with flammable gases, small pipe diameters favor the occurrence of detonations for combustible dusts [78, 79], but the distance required to reach the very high velocities is greater for dusts than for flammable gases which detonate with comparable violence.

As per Fig. 139, there seems to be a quasi-linear correlation between the maximum explosion pressure p_{max} and the maximum flame velocity v_{max} which is almost independent of the dust type for dust explosions initiated in pipelines with a flame jet ignition. The values deviate because of the different maximum explosion pressures of the dusts (see Fig. 88) and the superimposed oscillations of the reaction over time. Long pipelines with elbows and end flanges will show similarly high pressures as in Fig. 137 if such pipelines are attached to vessels which suffer dust explosions.

In summarizing, it can be stated that the course of combustion of dust/air mixtures in pipelines mainly depends upon the displacement effects with the resulting change in turbulence of the still unburned mixture ahead of the flame front. In particular, with flame jet ignition from attached vessels, such explosions may convert

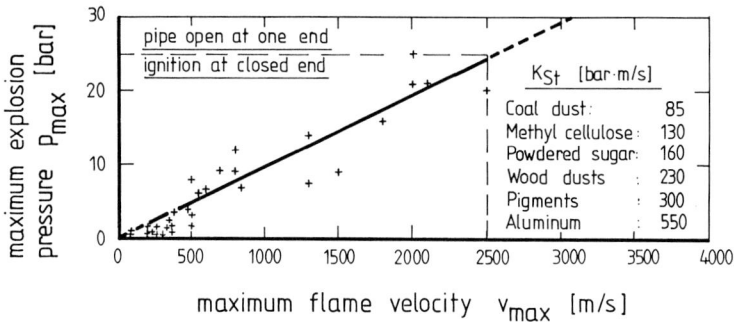

Fig. 139. Maximum explosion pressure p_{max} of dusts in a pipeline DN 400-mm, lp = 40-m as a function of the maximum flame velocity v_{max} with flame jet ignition

to detonations or quasi-detonations with substantial pressures. Small pipe diameters will favor a detonation. However, the pipe length required to reach the high constant velocity over a long stretch is longer for combustible dusts than for flammable gases. The maximum explosion pressure is generally higher in closed pipes than in open pipes, with extreme pressure values at connecting flanges or elbows.

5 Protective Measures Against the Occurrence and Effects of Dust Explosions

5.1 Preliminary Remarks

It was already mentioned (see sect. 2) that the phenomenon of dust explosions is not new. Reliable documented experience extends almost 200 years back in technical history. As long as organic chemistry is applied on an industrial scale, there are always unforeseen explosions (Figs. 1–4, 7, 140).

Pressure build-up and the resulting fires cause much damage to manufacturing equipment and buildings, along with fatalities. It is the task of safety technology to prevent such damage or limit its effects. It should not adjust the safety yardstick because of increased numbers of accidents, but rather take preventive action based on analysis. Explosion protection consists of the assessment of the explosion danger resulting from the handling of materials which may create dangerous and explosible atmospheres, plus the assessment of the effectiveness of protective measures for the prevention of such dangers. The risk of an accident must be reduced to a jus-

Fig. 140. Fire subsequent to a dust explosion

tifiable level. In the 1960s, the chemical industry worldwide suffered from dust explosions which could not be explained satisfactorily with the knowledge available at that time.

Therefore, in 1967, a few German and Swiss chemical companies, together with the Bergbau-Versuchsstrecke" in Dortmund-Derne, formed a collaborative circle called "Combustible Dusts". Its goal was to formulate possibilities for the prevention of dust explosions altogether or to limit the effects to an acceptable level. This was to be done in close collaboration with research institutes and the German Society of Engineers (VDI). In the last decade, all concerned contributed very successfully, as was reported in numerous VDI colloquia. Important insights have been shared relative to the practical application of explosion protection against dust explosions.

In 1976, the guideline for the prevention of the dangers of explosive atmospheres with examples was issued by the "Berufsgenossenschaft der Chemischen Industrie" and has been adopted. It is called in short Ex-RL [97]. It should facilitate the assessment for the user of whether a dust explosion danger exists in a given manufacturing installation. For clarification, the following questions are raised [81]:

– "Will there be an explosive atmosphere in the area of the installation to be analyzed or within the equipment?"
– "What is the amount of explosive atmosphere present or possible based on local or manufacturing conditions and where will it be generated?"
– "Is the amount of expected explosive atmosphere dangerous?"

If the questions are positively answered, then a decision has to be made with regard to the selection of the protective measure against the imminent dust explosion. The guideline distinguishes between three groups:

a) measures which will prevent or limit the formation of explosible atmospheres. Such a measure is often called "primary explosion protection".
b) measures which will prevent the ignition of explosible atmospheres and
c) measures which will limit the results of a dust explosion to a harmless level.

Measures a) and b) prevent the start of dust explosions and are summarized under the heading "preventive explosion protection" in the revised VDI-guideline 2263 "Dust fires and dust explosions: hazards-assessment-protective measures" [23]. Measure c), which limits the effects of dust explosions is classified as "explosion protection through design measures". The basis for the assessment of the scope of protective measures is the classification into zones, as per Ex-RL [80]. The zones reflect the probability of the existence of dangerous, explosible atmospheres. The definitions for combustible dusts are as follows:

Zone 10   includes locations in which dusts present a dangerous, explosible atmosphere intermittently or frequently.
Zone 11   includes locations in which an occasional dispersion of settled dust may be expected, forming a dangerous, explosible atmosphere.

The inside areas of equipment (e.g., mills, dryers, mixers, conveyors, ducts, silos,..) belong in general to zone 10.

Zone 11 covers the area surrounding dust-containing equipment. Because of leaks, dust may escape and settle in dangerous amounts (e.g., in a mill building where dust exits the mill and settles) [80, 81].

The following paragraph outlines the protective measures against dust explosions which are possible based on today's knowledge. A distinction will be made between preventive and design measures for the explosion protection [23]. For every case, it has to be determined whether the measure can be used by itself or in combination in order to reach the desired safety goal. In order to apply the described safety measures, not only the safety characteristics of the processed dusts have to be known (see sect. 4.2–4.5) but also the knowledge must be available pertaining to the start and the course of dust explosions.

5.2 Preventive Explosion Protection

5.2.1 Preliminary Remarks

A combustion reaction requires a fuel in the form of a combustible dust, oxygen from air, and sufficient energy for the ignition source, as shown in the hazard triangle depicted in Fig. 141. If one of the three conditions is missing there can be no dust explosion.

The triangle can be broken up by:

- preventing the formation of an explosible dust/air mixture
- replacing the oxygen with inert gas, working under vacuum or using inert dust or
- eliminating any efficient ignition source

All three measures are summarized under the heading "preventive explosion protection". Their appropriate application prevents a dust explosion from starting.

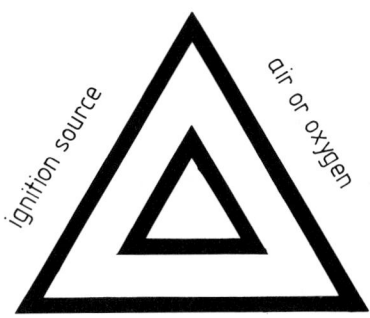

Fig. 141. Hazard triangle

5.2.2 Prevention of Explosible Dust/Air Mixtures

If it is possible to maintain the dust concentrations outside the explosible limits (ignition range), then dust explosions are prevented. Such a measure can be used alone or in conjunction with other safety measures. This depends upon the process technology and the safety characteristics of the dusts. In such a context, the lower explosible limit (LEL) is of particular importance (see sect. 4.3.1.4) (Fig. 142).

In individual cases, the limitation of the dust concentration in equipment or parts of the equipment can be used as the safety measure, provided a constant concentration can be guaranteed which is well below the LEL of the fine dust (M <63 μm) (e.g., room air exhaust or clean air downstream of filters). With time, some settling of dust has to be expected. The dispersion of the settled product may present an explosion danger which can be eliminated through frequent cleaning.

A simple overall consideration of, e.g., the amount of dust per vessel volume may not reflect the true concentration in all partial volumes due to the likely inhomogeneous dust distribution. The calculated average dust concentration for spray dryers shows that the LEL will not be reached. Experience, however, indicates that an explosible dust/air mixture has to be expected in the conical section of the dryer and in the equipment downstream from a dryer (cyclones, filters). In high-performance pneumatic conveyors, the upper explosible limit is in general exceeded due to the loading. Tests with a dust filter [81] have shown that a dust explosion with polyethylene fine dust (M = 26 μm; p_{max} = 7.5 bar; K_{St} = 104 bar.m/s) in the vented

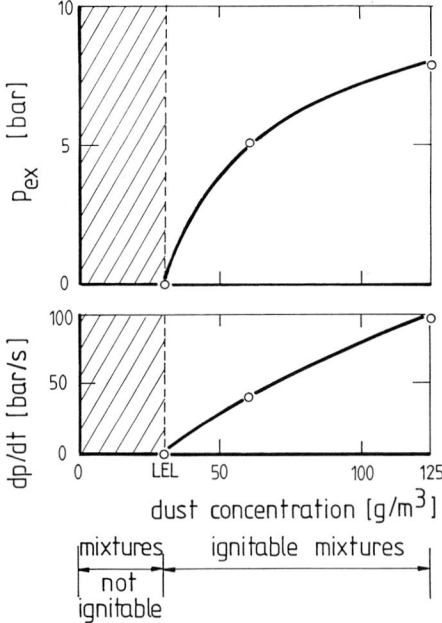

Fig. 142. Lycopodium: explosion pressure p_{ex} and rate of pressure rise dp/dt within the range of LEL (1-m^3 vessel)

5.2.3 Prevention of Dust Explosions by Using Inert Matter

filter housing will extinguish in the feed pipe DN 400 after 20–40-m provided that the dust concentration is at least 2–3 kg/m^3 or more. At start up or when the product enters the filter, cyclone, or silo the formation of an explosible dust mixture is possible. In such an operational phase, additional safety measures may be necessary, e. g., inerting (see sect. 5.2.3).

In order to prevent an explosible concentration de-dusting with a liquid may be advisable, for instance, with water or special oil [83].

5.2.3 Prevention of Dust Explosions by Using Inert Matter

5.2.3.1 Admixture of Nitrogen

5.2.3.1.1 Preliminary Remarks

In order to prevent dust explosions, the preventive measure inerting may be used. By introducing ample amounts of inert gas – mostly nitrogen – into the volume which needs to be protected (e. g., inside of containers, mills, silos) the volumetric oxygen content will be reduced to such a level that no ignition of the dust/air mixture will occur [23].

The application of such a safety measure calls for expert knowledge, mastery of the process and gas-tight equipment. By inerting the air of combustion, the ratio oxygen/nitrogen is changed in favor of nitrogen.

The lowest minimum ignition energy (LMIE) of combustible dust increases linearly in a semi-logarithmic plot (Fig. 143). The rapid decrease of the readiness for ignition of the mixture is naturally due to the nitrogen which as a ballast absorbs the heat released without participating in the combustion. Excess oxygen, however, flattens the curve [60].

With the change of the ignition source (extended capacitor discharge → pyrotechnic ignitors) a break occurs in the line for hard-to-ignite dusts (pea flour, melamine) in normal air because the pyrotechnic ignitor has a better efficiency due to its longer burn-time.

The oxygen concentration in nitrogen giving LMIE = 10^7 mJ (= 10,000 J) which is equivalent to the energy used for the determination of explosion indices, is called "limiting oxygen concentration (LOC)". The LOC of a dust/air/inert gas mixture is experimentally determined using varying dust concentrations and oxygen/nitrogen ratios. The oxygen concentration which will just not allow an explosion will be called the limiting concentration. It is a specific value for the dust and inert gas used and has to be determined experimentally due to the lack of any other correlation. For practical applications, a safety margin of generally 2 % is allowed, depending upon circumstances. Such a concentration is called "maximum allowable oxygen concentration [23]. This means that not all of the oxygen has to be displaced in order to prevent a dust explosion if the safety measure "inerting" is used.

Total exclusion of oxygen is only needed in case fires have to be prevented. It has to be pointed out that inerting is not an effective safety measure against thermal decompositions or deflagrations.

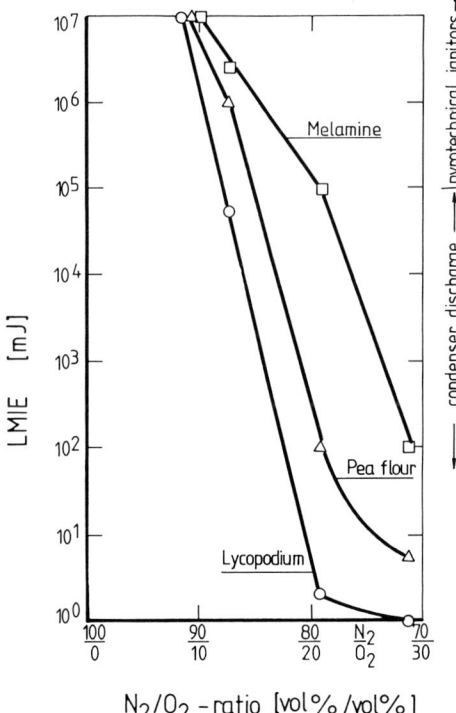

Fig. 143. Influence of the ratio oxygen/nitrogen in the air for combustion upon the lowest minimum ignition energy of combustible dusts

5.2.3.1.2 Combustible Dusts

As already stated (see sect. 4.3.1) it is customary to determine the explosion indices in a 1-m³ vessel or the 20-l laboratory apparatus. It was also pointed out (Fig. 54) that a reduction of the oxygen content through the addition of nitrogen to the air of combustion will reduce the explosible range. The oxygen reduction has to be greatest at the LEL in order to prevent a dust explosion. At the same time, the explosion indices are being reduced. They include:

– explosion pressure p_{ex}
– pressure rise dp/dt

which describe the course of a dust explosion in closed vessels as a function of the dust concentration. Figures 144 and 145 give a practical example from tests made in the large vessel as well as in the laboratory equipment.

In analyzing the effects of decreasing oxygen content upon the behavior of the optimum explosion characteristics

– maximum explosion pressure p_{max} and
– dust-specific K_{St}-value

5.2.3 Prevention of Dust Explosions by Using Inert Matter   131

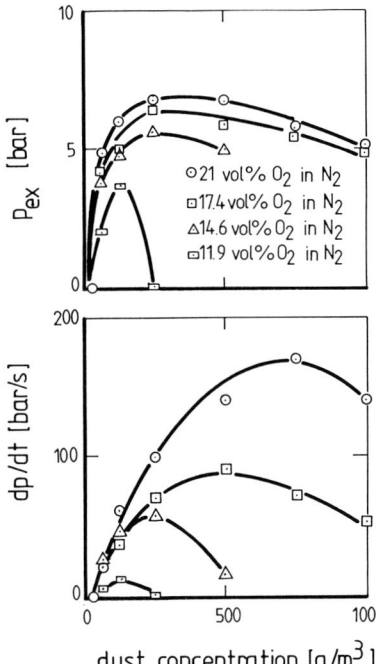

Fig. 144. 1-m³ vessel. Insecticide: Influence of oxygen content upon explosion indices

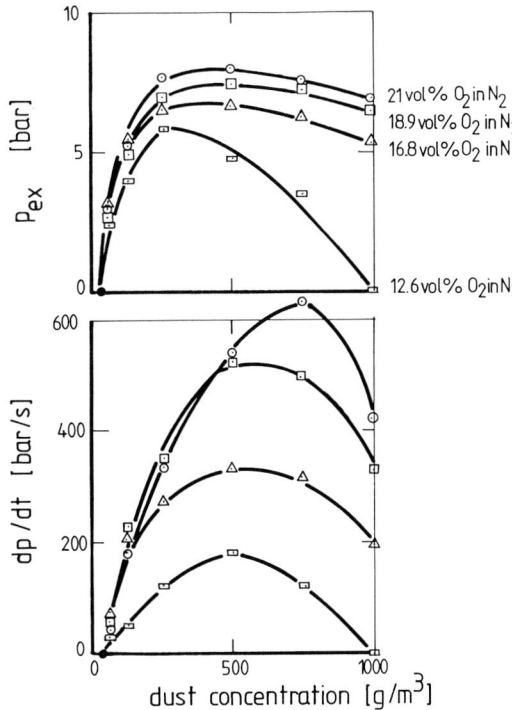

Fig. 145. 20-l laboratory apparatus. Insecticide: Influence of oxygen upon explosion indices

for either type of equipment the following can be stated (Fig. 146):

- For both equipment, the maximum explosion pressure p_{max} decreases slowly at first and then rapidly in the range of the limiting oxygen concentration. This is especially true for the larger vessel.
- The behavior of the K_{St}-value, however, noticeably depends upon the size of the test vessel. For the large vessel, there is a linear correlation with the oxygen content in nitrogen.

The intersection with the abscissa gives the limiting oxygen concentration. For the 20-l laboratory apparatus, however, the K_{St}-value approaches the abscissa asymptotically, rendering a much lower LOC than in the large vessel.

The reason for the sensitive behavior of the laboratory equipment in comparison with the larger vessel is the influence of the energy of the ignition source. The pyrotechnic ignitors which are used for the dust explosion testing have an energy content of 10,000 J and present a point source in the 1-m³ vessel. In the 20-l equipment, however, they affect the whole explosion volume once ignited. A number of combustible dusts were comparatively tested in both vessels [62]. The results gave the following correlation for the limiting oxygen concentrations (LOC):

LOC (1-m³ vessel) = 1.64 · LOC (20-l apparatus)

The LOC for the large vessel averages 64 % higher than the laboratory equipment. LOC values determined in the laboratory equipment have to be adjusted in line

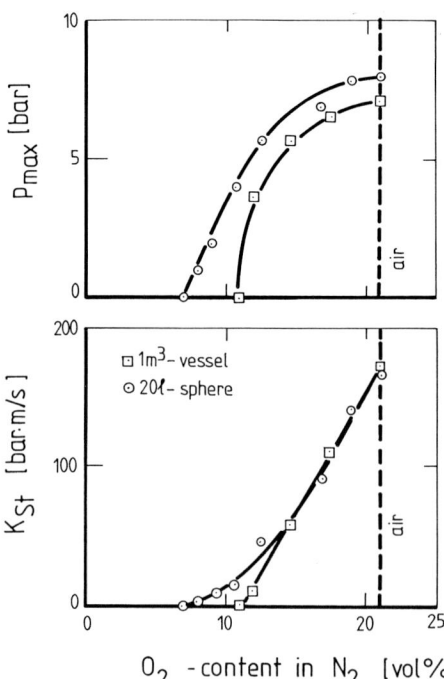

Fig. 146. Insecticide: Influence of the oxygen content in nitrogen upon the maximum explosion pressure and K_{St}-value in the larger (1-m³) and smaller laboratory equipment

5.2.3 Prevention of Dust Explosions by Using Inert Matter

with the above equation for large vessels before they are suitable for practical application. References [55, 56] accept only equipment for dust testing which renders the same safety characteristics for airborne dusts as the standard 1-m³ vessel.

The LOC for combustible dusts has to be determined experimentally from case to case due to the lack of a correlation. But there are certain tendencies to be seen [63].

- the LOC is higher, the weaker the dust reacts (Fig. 147)
- the LOC is higher, the higher the ignition temperature in the BAM-oven is (Fig. 148).

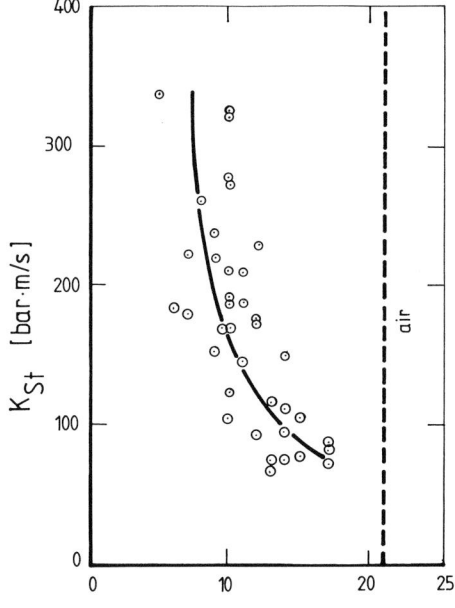

Fig. 147. Correlation of K_{St}-value with the limiting oxygen concentration (LOC) (1-m³ vessel)

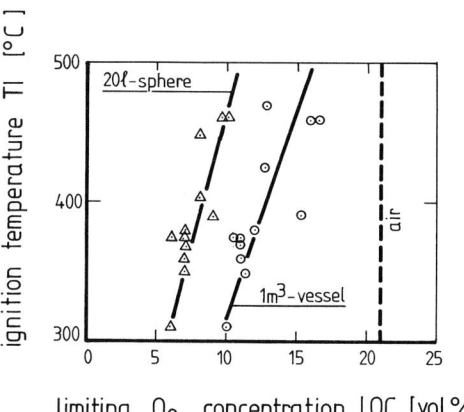

Fig. 148. Correlation of ignition temperature with limiting oxygen concentration (LOC) (1-m³ vessel)

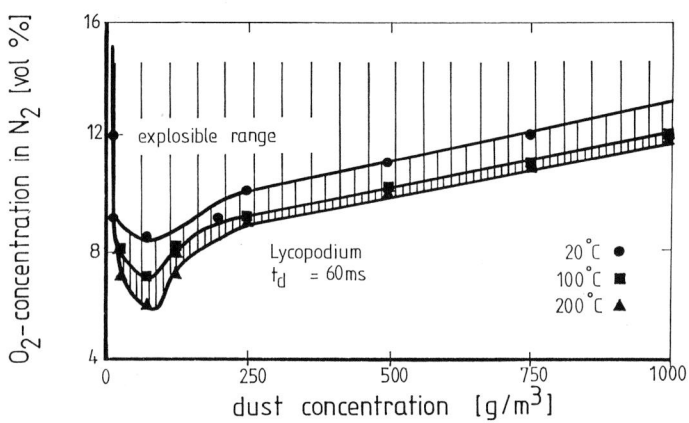

Fig. 149. Lycopodium: Influence of the limiting oxygen content in nitrogen and the temperature upon the dust concentration (20-l laboratory apparatus)

In industrial practice, the inerting safety measure is often applied at above room temperature for grinding and drying operations. Glarner [59] systematically studied the influence of temperature upon LOC in a heated 20-l laboratory equipment (Fig. 70).

Figure 149 shows the influence the oxygen concentration and the temperature have upon the explosible range of Lycopodium. Increasing temperature expands

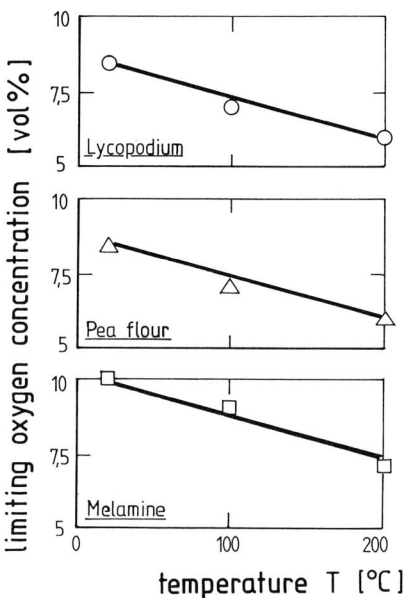

Fig. 150. Influence of the temperature upon the limiting oxygen concentration (LOC) of combustible dusts

5.2.3 Prevention of Dust Explosions by Using Inert Matter

the explosible range and reduces the LOC [63]. Because of the resulting linear correlation (Fig. 150), the average decrease amounts to 1.8 vol% per 100 °C temperature increase.

Wiemann [84] investigated the temperature dependence of the LOC for various dusts in a heated 1-m^3 vessel. He also found a linear correlation, with an average decrease of 1.4 vol% per 100 °C temperature increase. From this, it can be concluded that the influence of the vessel size V<1-m^3 has to be considered and corrected for in the determination of the LOC but that the temperature dependency is not affected within the range of accuracy. The LOC is dust-specific and, in general, requires an experimental determination. Its is not only dependent upon the dust type but also upon the particle size. For a few combustible fine dusts, the LOC at room temperature is summarized in Table 16, with nitrogen as the inerting medium.

Table 16. Combustible dusts: limiting oxygen concentration (LOC) for inerting with nitrogen (1-m^3 vessel, E = 10,000 J, room temperature)

Dust type	M [μm]	LOC [vol%]
Pea flour	25	15.5
Cadmium laurite	<63	14
Hard coal	17	14
Barium stearate	<63	13
Rye flour	29	13
Brown coal	63	12
Soot	13	12
Organic pigment	<10	12
Herbicide	10	12
Cadmium stearate	<63	12
Calcium stearate	<63	12
Wheat flour	60	11
Polyacrylonitrile	26	11
Cellulose	22	10.5
Wood	27	10
Resin	63	10
Methylcellulose	70	10
Polyethylene (HDPE)	26	10
Dibenzoylperoxide	59	10
Beta-naphthol	<30	9.5
Bisphenol A	34	9.5
Corn starch	17	9
Sulfur	30	7
Paraformaldehyde	23	6
Aluminum	22	5

Table 17. Brown coal: Influence of the type of inert gas/vapor upon the limiting oxygen concentration (LOC) (1- m³ vessel, $E = 10{,}000$ J, $T = 150°C$)

Inert gas/vapor	LOC [vol%]
Nitrogen	11.0
Water vapor	12.4
Carbon dioxide	13.0

For most organic dusts, a reduction of the oxygen content by half (relative to air) results in a complete inerting. If the maximum allowed oxygen concentration in nitrogen is maintained at 8 vol%, then no dust explosion will be possible. But there are exceptions, e. g., beta-naphthol, paraformaldehyde. Wiemann [84] also investigated the influence of the type of inert gas or vapor at elevated temperatures upon the LOC for brown coal (Table 17).

The effectiveness of water vapor is between that of nitrogen and carbon dioxide.

If "inerting" is used a a safety measure in practice for the protection of vessels and equipment, the LOC has to be monitored in such a way that it will not be exceeded.

This may be accomplished by continuous or intermittent measuring. Special attention has to be given to the following [85]:

- selection of a suitable parameter for monitoring (oxygen concentration, inert gas flow)
- selection of suitable instrumentation
  (e. g., principle of measurement, sensitivity, accuracy, response, response delay due to the distance sampling point – measuring point)
- selection of a suitable sampling point considering the flow pattern, which must reflect the worst condition for the oxygen concentration. Multiple sampling points may be necessary.

The proper measures have to be defined in case of an upset:

- Shutting down the equipment in case the highest allowable oxygen concentration is exceeded or a higher flow of inert gas is called for than required for proper inerting. In general, shut-down should be automatic.
- Selection of the proper level for an alarm which, once exceeded, will trigger counter-measures (automatic, mechanical).

If the measuring is done intermittently, it has to be ensured that:
- the system is completely closed with well-defined reproducible gas flows (danger: cleaning ports)
- the necessary test period was long enough to arrive at safe operating conditions for inerting

- if the system is subsequently modified, the operating conditions for inerting will be checked and reset if necessary
- the degree of inerting will be known for all operating conditions and affected pieces of equipment
- an alarm will be actuated in case the supply of inert gas fails.

5.2.3.1.3 Hybrid Mixtures

In case hybrid mixtures, consisting of combustible dust and flammable gases, have to be inerted, the behavior of the range of ignition of the flammable gas has to be known at reduced oxygen levels [63]. Such a reduction is, of course, accomplished by adding nitrogen to the air of combustion. The behavior of propane is shown in Fig. 151 for the case of ignition:

1) in a quiescent state and
2) the same turbulence used for dust explosion tests.

A reduction in oxygen content will not change the lower explosive limit (LEL) but will markedly reduce the upper explosive limit (UEL), similar to combustible dusts. The state of the mixture has a slight influence upon the ignition range but not upon the limiting oxygen concentration (LOC), which is 10 vol%. However, an additional consideration involves the influences upon the limiting concentration from the potent ignition source ($E = 10,000$ J) used for dust testing. The LOC is markedly higher at 11.8 vol% if the standard ignition source for flammable gas testing is used, namely the continous induction spark with an energy $E = 10$ J.

Figure 152 shows the correlation of the explosion characteristics with the degree of inerting with nitrogen for pea flour, propane, and their hybrid mixtures. The LOC for dust is at 15.5 vol%, markedly higher than the one for the flammable gas, which is 10 vol%.

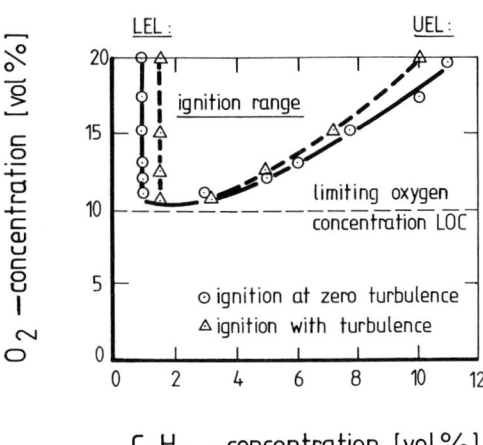

Fig. 151. Propane: Influence of oxygen content in nitrogen upon the flammable range of mixtures ignited in either a quiescent or turbulent state

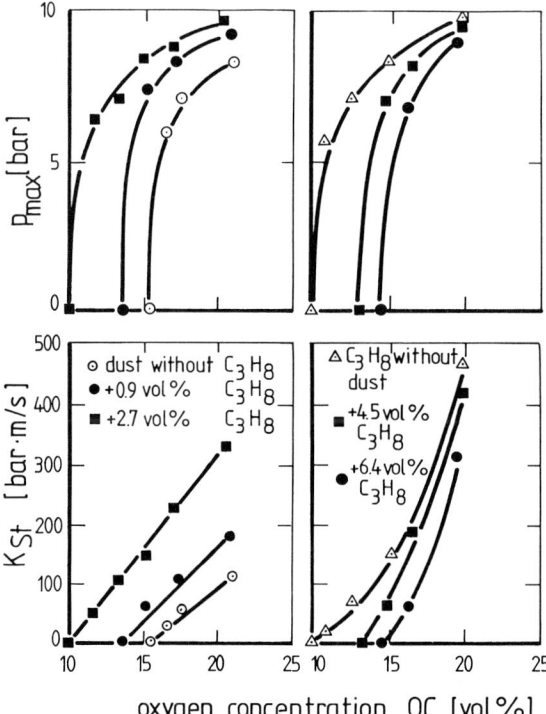

Fig. 152. Inerting of the hybrid mixture pea flour and propane using nitrogen addition to the combustion air (1-m³ vessel, E = 10,000 J)

The reduction of the oxygen content causes a slight reduction in the maximum explosion pressure p_{max}, at first, independent of the propane concentration, and then gives a drastic change at the LOC. It influences the dust-specific K_{St}-value linearly, in the same fashion as with dust alone, if the propane content is near its LEL.

The K_{St}-value approaches the abscissa asymptotically with a high gas content similar to propane alone.

In general, hybrid mixtures can be ignited at oxygen concentrations which will not allow pea flour to explode. The LOC of the mixture moves towards the one for propane alone and then away from it, for higher flammable gas concentrations. Such a statement is generally valid for any combustible dust (Fig. 153) and flammable gas (Fig. 154). In order to allow comparisons in the latter case, the concentrations are referenced to the optimum concentration for the explosion indices. This serves as the standard.

In summary, it can be stated:

For hybrid mixtures consisting of combustible dust and flammable gases or vapors, the LOC is given by the fuel with the lowest limiting value. For flammable gases, the following correlation exists for the LOC determined in the 1-m³ vessel and the 20-l laboratory apparatus [63]:

LOC (1-m³ vessel) = 1.32 · LOC (20-l laboratory app.)

5.2.3 Prevention of Dust Explosions by Using Inert Matter 139

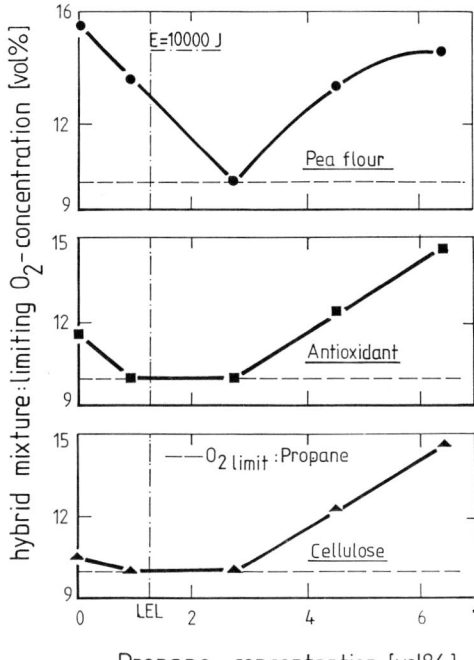

Fig. 153. Hybrid mixtures of combustible dusts with propane. Correlation of limiting oxygen concentration with propane concentration
(1-m^3 vessel, E = 10,000 J)

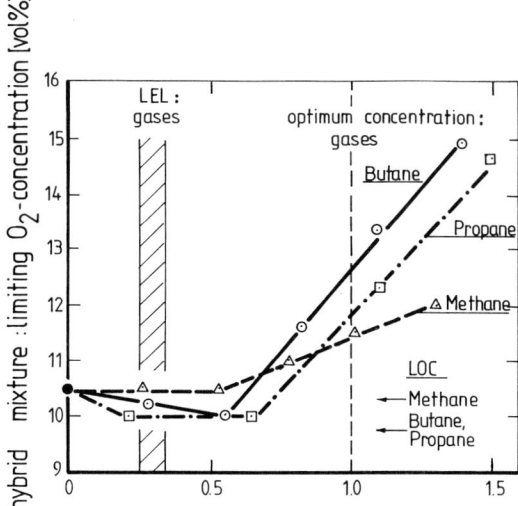

Fig. 154. Hybrid mixtures of cellulose with flammable gases. Correlation of limiting oxygen concentration with flammable gas concentration
(1-m^3 vessel, E = 10,000 J)

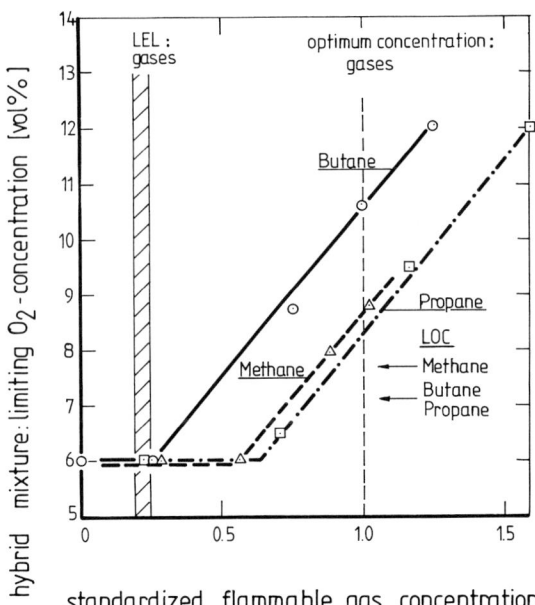

Fig. 155. Hybrid mixtures. Cellulose with flammable gases. Correlation of limiting oxygen concentration with flammable gas concentration (20-l laboratory apparatus, E = 10,000 J)

On the average, the LOC in the large equipment is 32% higher than in the 20-l laboratory apparatus. The size of the test vessel influences the results for flammable gases only half as much as in the case of combustible dusts (see sect. 5.2.3.1.2). As a consequence, the LOC for, e.g., cellulose with 10.5 vol% in the large equipment is above, and the 6.0 vol% in the laboratory equipment markedly below the one for the flammable gases butane, propane, and methane. But this case again shows (Fig. 155) that the LOC is determined by the fuel with the lowest limit.

Otherwise, the statements made in sect. 5.2.3.1.2, for the monitoring of the highest allowable oxygen concentration for combustible dusts are also valid for hybrid mixtures.

5.2.3.1.4 Use of Vacuum

It was already mentioned (Fig. 79) that the explosion indices of combustible dusts (maximum explosion pressure p_{max}, dust-specific K_{St}-value) in closed vessels are proportional to the initial pressure p_i, i.e., the starting pressure for the explosion. By reducing the starting pressure below atmospheric, one can accomplish the following:

– no dust explosion will occur (this is generally the case for initial pressures $p_i \leq 50$ mbar) or
– the maximum explosion pressure remains below atmospheric pressure (for organic dusts, this will be the case at initial pressures $p_i \leq 0.1$ bar).

5.2.3 Prevention of Dust Explosions by Using Inert Matter

At the same time, the lowest minimum ignition energy increases (Fig. 115). The larger the negative pressure, the higher the safety margin. Such a safety measure against the start of a dust explosion is often used for vacuum-rotary dryers. The vacuum has to be monitored with instruments and in case of failure (e.g., air entrainment) be replaced with another safety measure (e.g., inerting or preventing of ignition source through shut-off).

5.2.3.1.5 Admixture of Solids

Combustible dusts may be inerted through the additional of inert powdery solids [63]. This has to happen in such a fashion that the fuel is intensively mixed with the inert powder.

If, e.g., pea flour is inerted with an extinguishing powder having ammonium phosphate as a basis (Tropolar), then the lowest minimum ignition energy (LMIE) of the combustible dust/inert powder mixture will increase rapidly with increased inert powder content (Fig. 156). Here, the correlation is similar to the ignition behavior of combustible dusts inerted with nitrogen (Fig. 143). Therefore, that inert powder concentration in the whole mixture which requires lowest minimum ignition energy LMIE = 10^4 J is defined as the limiting inert powder concentration. In Fig. 156, this would correspond to 75 % Tropolar in the combustible dust/inert powder mixture. As there exists no general correlation for inerting with solids, the so-called limiting concentration has to be determined for every application.

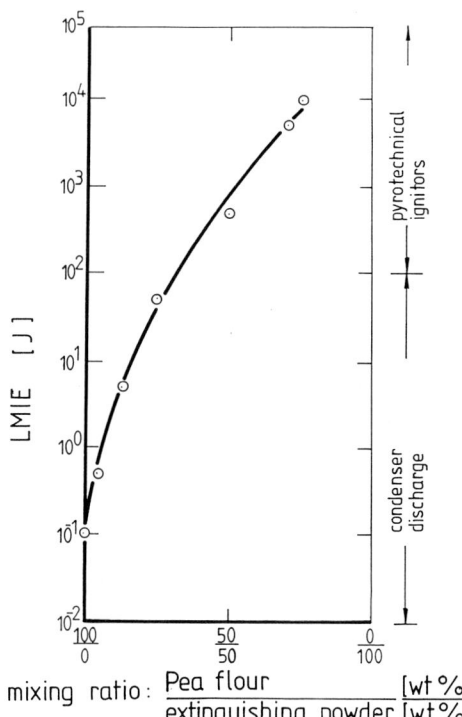

Fig. 156. Influence of the mixing ratio pea flour/Tropolar upon the lowest minimum ignition energy (LMIE) of the mixture

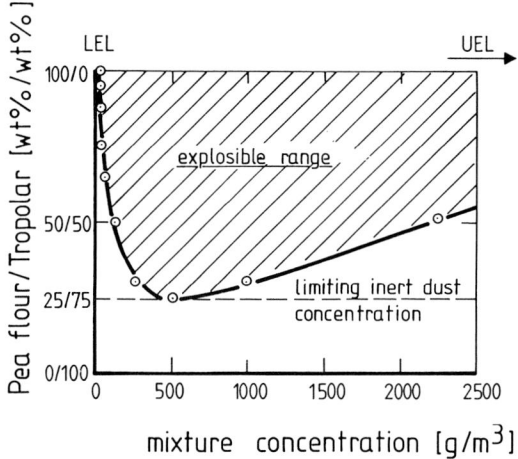

Fig. 157. Influence of the mixing ratio pea flour/inert powder (Tropolar) upon the ignition range of the mixture (20-l laboratory apparatus, E = 10,000 J)

Figure 157 shows the influence upon the ignition range of the admixture of Tropolar to pea flour. There is only a minor influence upon the LEL, but a pronounced limitation of the UEL can be observed, similar to the inerting with nitrogen. At the same time, the explosion indices (maximum explosion pressure p_{max}, dust specific K_{St}-value) decrease and reach zero at 75 wt% Tropolar (Fig. 158).

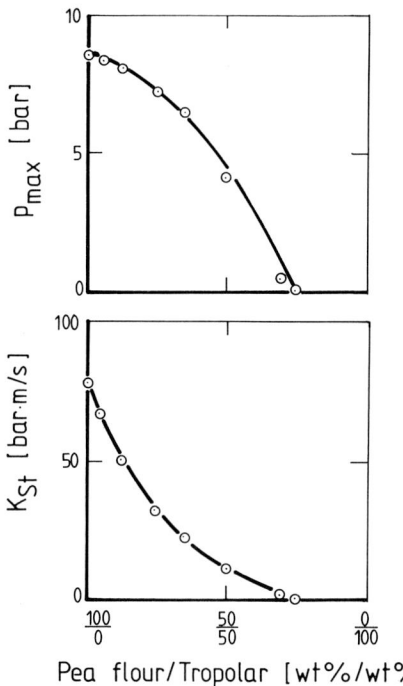

Fig. 158. Influence of the mixing ratio pea flour/inert powder (Tropolar) upon the explosion indices (20-l laboratory apparatus, E = 10,000 J)

5.2.3 Prevention of Dust Explosions by Using Inert Matter

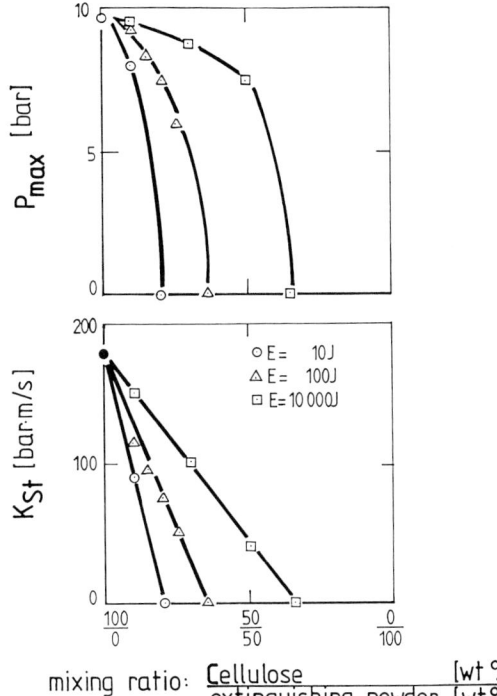

Fig. 159. Influence of the ignition energy (IE) upon the inerting of cellulose with inert dust (Tropolar) (1-m³ vessel, E = 10,000 J)

Another parameter which influences the limiting inert concentration is the ignition energy used (Fig. 159). Results from tests in the 1-m³ vessel indicate that lowering the ignition energy also lowers the limiting concentration for inert powder admixture. If not so potent ignition sources are actually expected in the equipment or apparatus to be protected, then the admixture of only approximately 20 wt% of inert powder (Tropolar) will suffice to inert a cellulose mixture.

There are no comparative test results available for the limiting inert powder concentration in the 1-m³ vessel and the 20-l laboratory apparatus. But the determination of this value means also to determine an explosion characteristic in both apparatus which does not influence the composition of the air of combustion. As concurring characteristics have been proven for the large and laboratory test equipment (Fig. 67), it can be anticipated that the limiting inert powder concentration will not fluctuate in the same major fashion as the LOC. Analogous to inert gases, inert solids have a varying effectiveness when inerting combustible dusts as shown in Table 18.

In conclusion, Table 19 shows the required minimum concentration of inert powders necessary to completely inert combustible dusts subjected to the usual potent ignition source for dust testing:

Table 18. Effectiveness of various inert powders for inerting combustible dusts (1-m³ vessel, E = 10,000 J)

Combustible dust	Effectiveness of inert powder	
	favorable	unfavorable
Coal dust	$NH_4H_2PO_4$	$KHCO_3$
Sugar	$NaHCO_3$	KCl
Dextrin	$NaHCO_3$	KCl
Organic pigment	$NH_4H_2PO_4$	$NaCl$
Aluminum	$NaHCO_3$	$NaCl$

Table 19. Inerting of combustible dusts by admixing inert powders (E = 10,000 J)

Combustible dust	M [μm]	Inert powder	M [μm]	Required minimum concentration for inerting [wt%]
Pea flour	25	$NH_4H_2PO_4$	29	75
Methyl cellulose	70	$CaSO_4$	<15	70
Cellulose	22	$NH_4H_2PO_4$	29	65
Hard coal (bituminous)	20	$CaCO_3$	14	65
Hard coal (bituminous)	20	$NaHCO_3$	35	65
Organic pigment	<10	$NH_4H_2PO_4$	29	65
Aluminum	<10	$NaHCO_3$	35	65
Dextrin	<63	$KHCO_3$	25	55
Sugar	30	$NaHCO_3$	35	50

Generally, more than 50 wt% of inert powder has to be admixed for inerting. However, as shown in Fig. 159, markedly lower concentrations of inert powders are needed if less potent ignition sources are anticipated in industrial practice.

5.2.4 Prevention of Effective Ignition Sources

5.2.4.1 Preliminary Remarks

Dust explosions can be prevented by avoiding effective ignition sources which are capable of igniting dust/air mixtures because of their characteristics, e. g., energy, temperature, and duration. In addition to the normal course of operations, operational upsets have to be considered which may generate ignition sources [22]. The application of the safety measure "prevention of effective ignition sources" has to be justified and assumes sound knowledge.

The following distinction has to be made:

- trivial ignition sources (e.g., unauthorized smoking, welding, cutting, using open flames)
- ignition sources inherent in the process or operational upset (e.g., foreign particles, pin breakage in mills, glowing particle nests).

Trivial ignition sources can positively be excluded in a well-managed modern facility through administrative measures (e.g., permits). Inherent operational ignition sources can be prevented through the use of dust explosion-proof electrical equipment, adhering to temperatures given by safety considerations, sound bonding and grounding for electrostatic purposes, and avoidance of mechanical drives with high rpm or power. Once both ignition categories can be excluded, then, based on today's knowledge, such a safety measure can be considered sufficient to prevent the start of a dust explosion. Special precaution is necessary for hybrid mixtures and dusts having a low LMIE < 10 mJ. Every installation has to be analyzed with regard to potential ignition sources and whether they can be avoided with a sufficient margin of safety. The guideline for explosion protection [80] lists 13 types of ignition sources. In the following, the most important ignition sources for combustible dusts will be characterized.

5.2.4.2 Mechanically Generated Sparks

Mechanical sparks, which are generated in grinding and milling equipment with rotating components having high peripheral velocities, are considered to have caused 30% of the dust explosions in industrial practice. An analysis of industrial mishaps has led to the following theory with regard to the ignition capabilities of sparks from rotating steel parts in dust/air mixtures. The conditions for the relative velocity of components are as follows:

$v < 1$ m/s: there is no danger for ignition
$v < 1$–10 m/s: every case has to be judged separately considering the product and material-specific characteristics
$v > 10$ m/s: in every case there is danger for ignition.

The assumption that there is no ignition danger at low relative velocities has been confirmed by recent test results [85]. In these tests, grinding sparks are generated by quickly contacting, e.g., a steel pin with a vitrified type ceramic wheel. Friction sparks and hot surfaces, however, are generated by rubbing steel pins against steel wheels over an extended time. Figure 160 shows the boundaries for the generation of grinding and friction sparks, as well as hot surfaces as a function of the thrust p_A and the peripheral wheel velocity. It illustrates that at velocities $v \leq 1$ m/s no ignition is to be expected unless exceptionally high thrust will generate steel sparks or hot steel surfaces. The materials of construction listed in Fig. 160 are referenced to the "steel nomenclature" [87] as follows (Table 20).

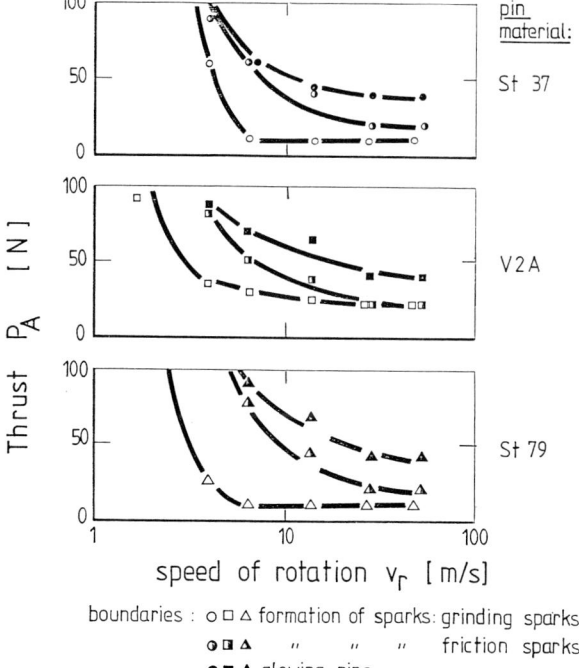

Fig. 160. Boundaries for the generation of grinding and friction sparks as well as hot surfaces (pin diameter 4 mm; materials for grinding wheel: V2A, St 37, St 79)

Table 20. Materials of construction

Marking	Material reference No [87]
V2A	1.4435
St 37	1.0037
St 79	1.1663

In order to compare the ignition capabilities of the different types of mechanically generated sparks in fuel/air mixtures, an electrical equivalent energy (EE) is assigned. EE is defined as the energy released through the discharge of a capacitor over an extended time period (see sect. 4.4.1). which has the same ignition capability as a given mechanically generated spark [88]. The procedure will be explained with the example of the flintstone friction spark (Fig. 161).

Figure 162 shows for propane the correlation of minimum ignition energy with the flammable gas concentration. Included also is the range of ignition for the flintstone friction spark. For such a spark there exists an electrical equivalent energy $EE = 4 \cdot 10^3$ mJ = 4 J for a below stoichiometric gas concentration. Above the stoi-

5.2.4 Prevention of Effective Ignition Sources

Fig. 161. Flintstone friction spark

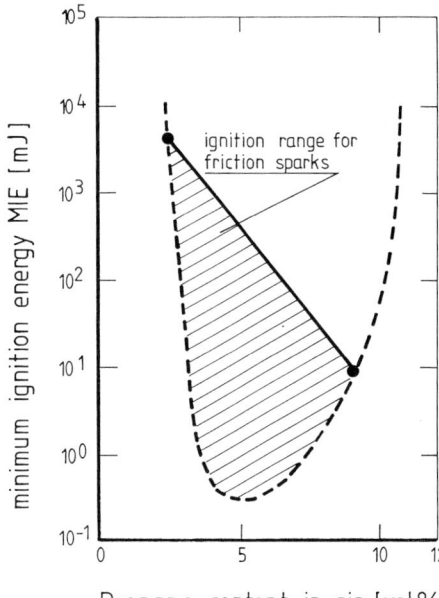

Fig. 162. Comparison of the minimum ignition energy of propane with the range of ignition of a flintstone friction spark

chiometric concentration EE = 9 mJ. The reduction in ignition effectiveness of the mechanical spark in rich fuels can be explained by the reduced oxygen content which impairs the spark combustion. Once the tests are made with other flammable gases one notices (Fig. 163) that the electrical equivalent energy decreases with

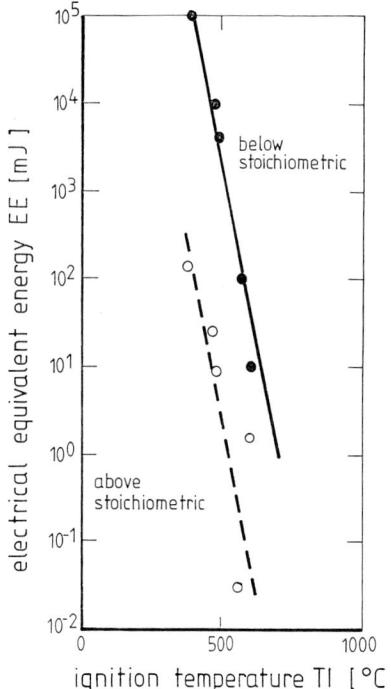

Fig. 163. Flintstone friction spark/flammable gases; correlation of electrical equivalent energy with ignition temperature

increased ignition temperature. In a semi-logarithmic plot one obtains two ignition lines. Of special interest for practical applications is the one for the range below stoichiometric concentrations. It is apparent that the electrical minimum ignition energy as well as the ignition temperature of the flammable gas influences the ignition effectiveness of a mechanically generated spark.

In the case where a flintstone friction spark is exposed to a dust/air mixture (Fig. 164), the connecting line limiting the ignition conditions will run horizontally since the dispersed dust will not influence the oxygen concentration, contrary to the behavior of flammable gases. The effectiveness of a flintstone friction spark in a lycopodium/air mixture at medium turbulence ($t_d = 90$ ms) is the same as an extended capacitor discharge, i.g., $E = EE = 100$ mJ. Mechanically generated sparks have, analogous to electrical sparks (Fig. 105), the best ignition effectiveness in low turbulent mixtures at markedly higher ignition delay times [88]. The energies stated in the following outlines are based on such a low turbulence. Figure 165 shows the correlation of electrical equivalent energy (EE) with the ignition temperature for a number of flammable gases and combustible dusts. A single line can be drawn as the boundary, which is independent of the type of fuel within the range of accuracy. Such an observation is also true for all other types of mechanical sparks.

5.2.4 Prevention of Effective Ignition Sources

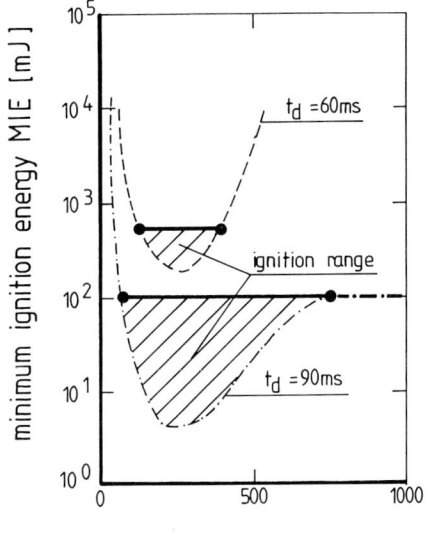

Fig. 164. Comparison of the minimum ignition energy of Lycopodium with the ignition range of a flintstone friction spark

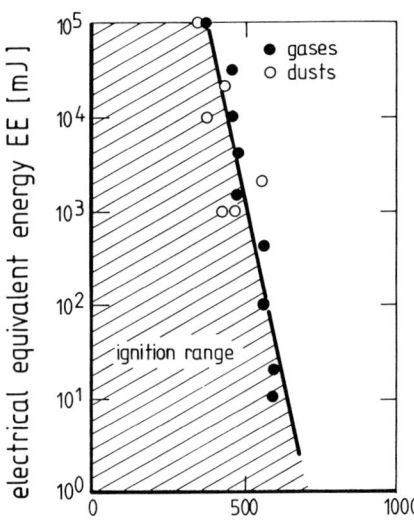

Fig. 165. Boundary line for ignition of fuel/air mixtures with flintstone friction sparks

150   5 Protective Measures Against the Occurrence and Effects of Dust Explosions

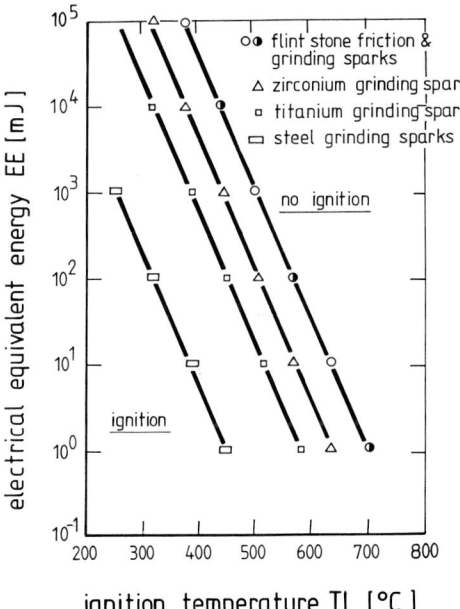

Fig. 166. Grinding sparks: correlation of electrical equivalent energy (EE) and ignition temperture TI

Grinding sparks occur if materials of construction briefly touch an, e.g., vitrified type ceramic wheel (grinding time t_g = 20–50 ms). For such a case (Fig. 166), parallel lines can be drawn, within the range of accuracy, that give the boundary for ignition, which depends upon the type of spark-producing material. Ignition of the combustible dust/air mixture is always to be expected if the minimum ignition energy is below the boundary, and no ignition, if it is above the boundary line. It is also obvious that dust/air mixtures with a low ignition temperature (e.g., TI = 300°C) will ignite even if they have a high LMIE. If the ignition temperature, however, is high (e.g., TI = 600°C), then only mixtures can be ignited having a very low value for the LMIE. Grinding sparks which are generated by grinding steel against steel [85] will, based on today's knowledge, only ignite mixtures of dust having ignition temperatures TI <300°C and LMIE <10 mJ. Sulfur, for instance, is such a substance.

Impact sparks have a very different ignition effectiveness, which depends upon the materials used to generate the sparks, as shown in Figs. 167 and 168. This is also reflected by the course of the corresponding ignition boundary lines (Fig. 169) [88].

Based on the results presented so far, the ignition effectiveness of mechanically generated sparks decreases as follows (Figs. 166–169):

- flintstone, friction and grinding sparks
- zirconium, grinding sparks
- titanium grinding and impact sparks
- steel grinding sparks and
- aluminum/rust impact sparks.

5.2.4 Prevention of Effective Ignition Sources

Fig. 167. Aluminum/rust impact spark

Fig. 168. Titanium/rust impact spark

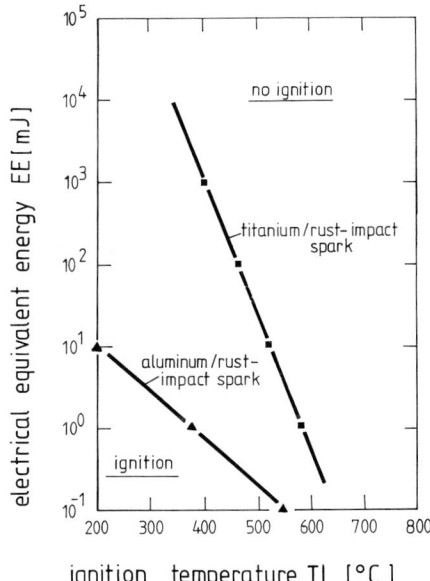

Fig 169. Impact sparks: correlation of electrical equivalent energy (EE) with ignition temperature TI

Steel grinding sparks (Fig. 170) have a relatively poor ignition effectiveness among the examples given (Fig. 166). With an ignition temperature TI = 400 °C as a reference, they are only capable of igniting dust mixtures having LMIE <10 mJ. Such behavior does not explain the previously stated frequent occurrence of dust explosions initiated by mechanically generated sparks in industrial practice. If steel is rubbed against steel (Table 20) for a longer duration, e.g., rubbing times $t_r = 0.5$–2.0 s, then friction sparks are generated (Fig. 171) which are much more ignition-efficient (Fig. 172). At the same ignition temperature, dusts will be ignited which have LMIE <100 mJ [86]. The likely reason for such behavior is the higher starting temperature of the sparks due to the heating-up of the pin. Therefore, the occurrence of the above-mentioned ignition source in practice is more likely.

In summarizing, it can be stated than the type of spark-producing material, together with the ignition temperature and the LMIE requirement, will determine whether an ignition of the dust/air mixture has to be anticipated from grinding, friction, or impact sparks.

5.2.4 Prevention of Effective Ignition Sources

Fig. 170. Steel grinding spark

Fig. 171. Steel friction spark

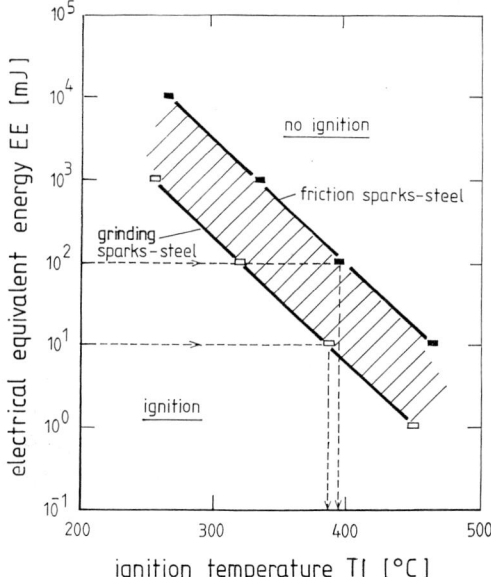

Fig. 172. Steel grinding and friction sparks: Correlation of electrical equivalent energy (EE) with ignition temperature TI

5.2.5 Hot Surfaces/Autoignition

Hot surfaces can, through direct contact, initiate dust explosions as well as ignite dust layers. Operational hot surfaces are expected with surfaces of hot equipment, heaters, dryers, steam piping, or electrical equipment [23, 80].

Besides the hot surfaces which are given by operational conditions, mechanical friction may cause dangerous temperatures in the exposed areas, e. g., in motors, fans, mechanical conveyors, mills, mixers, journal bearings, and rolling-contact bearings. This is also true for pieces of tramp metal wedged in between an agitator and a mixer wall. Newer findings [85] indicate that surface temperatures of at least 1200°C are needed to ignite dust/air mixtures through steel pins rubbing against steel discs (Table 20) (Fig. 173).

In using pins with a diameter of 4 mm or 6 mm, respectively, said temperature is reached after a rubbing time $t_r = 1.5-3.5$ s. The materials V2A and St 37 heat up most readily and are, therefore, most ignition efficient. The ignition capability of such a heated surface depends upon the optimum temperature T_{opt} reached, the diameter of the rubbed spot D_R, and the length of zone of discolorations, plus the temperature difference ΔT at the rubbed spot. The larger the heated surface, the easier the ignition of the dust/air mixture (Fig. 174). Whether ignition occurs depends additionally upon the ignition temperature of the combustible dust (see sect. 4.4.2.1).

As shown in Fig. 174, 4-mm steel pins will ignite dust/air mixtures having ignition temperatures up to TI = 420°C. An additional influence of the material of con-

5.2.5 Hot Surfaces / Autoignition

Fig. 173. Hot surface of a structural steel pin after rubbing against a structural steel disc

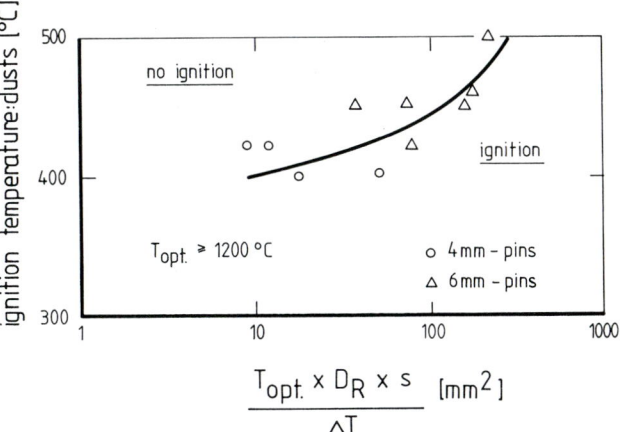

Fig. 174. Correlation of heated pin surface with ignition temperature of combustible dusts

struction upon the course of ignition could not be observed within the range of measuring accuracy.

In addition, auto-ignition may occur in dust-piles (see sect. 4.2.5). The resulting glowing particle nests or fires may themselves become the ignition source for dust/air mixtures. The question whether a glowing particle nest is actually capable of igniting a dust/air mixture cannot be answered at present. Experience indicates that in certain cases (coal dust) this is very hard to accomplish; in others (e.g., certain pigments) it is easily done.

5.2.6 Static Electricity

Charged static electricity upon its discharge may present an ignition danger for explosible dust/air mixtures [88–91]. Such an ignition source is responsible for approximately 9% of the explosion incidents in practice (Fig. 9).

Spark discharges always occur between two electrodes, e.g., an insulated charged metallic part and a grounded metallic part. The charging can be done either through direct contact with, for example, a charged product or through inductance, i.e., charge separation in an electrical field. Practical examples include isolated support cages for filter bags or a sieve or perforated plate with a non-conductive gasket [91].

The danger for ignition exists once the energy calculated from the capacitance C and voltage U as $E = 1/2\ CU^2$ is larger than the LMIE of a dust. One has to consider that a true spark discharge from a capacitor is less capable of igniting a mixture than an extended discharge (see sect. 4.4.1.2). But there are a few easily ignitable dusts which have a LMIE which is independent of the existence or non-ex-

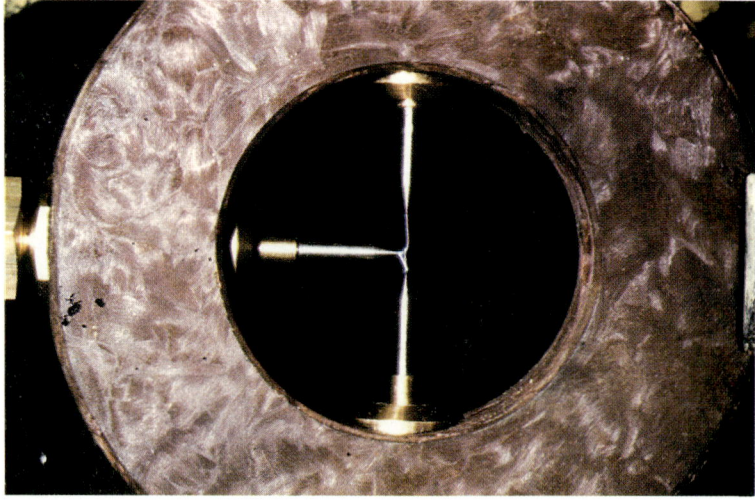

Fig. 175. Extended capacitor discharge at an energy = 1 mJ

5.2.6 Static Electricity

istence of an inductance in the discharge circuit. Spark discharges can be prevented in practice if all conductive parts which can be charged [23] are bonded and grounded electrostatically (resistance not larger than $10^6 \, \Omega$).

Brush discharges (Fig. 176) are generated in numerous processes. They can be caused by highly charged non-conductors, e.g., foils, filter media, plastic pipes, and deposited or dispersed dust.

Ignition tests with brush discharges in flammable gas/nitrogen/air mixtures have shown that they are as ignition-capable as extended capacitor discharges with an energy of a few mJ. Therefore, one should expect that brush discharges would ignite dust/air mixtures with a very low LMIE requirement which is independent of an inductance (LMIE ≤ 1 mJ). However, despite intensive testing, such proof was not obtained experimentally, which leads to the assumption that this type of discharge has a different behavior in dust/air mixtures [92] than in flammable gas/air mixtures. Since brush discharges present no ignition source for dust/air mixtures as per current experience, no additional measures are necessary to dissipate such charges, e.g., from filter cloths or the product proper. For hybrid mixtures, however, protective measures are needed, such as preventing large dust clouds, using conductive filter cloths, and possibly adopting design measures.

Fig. 176. Brush discharge

Fig. 177. Propagating brush discharges

Propagating brush discharges (Fig. 177) occur through separating or intensively rubbing non-conductive layers (thickness <8 mm) which are attached to a conductive support e.g., metallic.

Such a discharge has been observed inside air jet mills with non-conductive liners, cyclones, and chutes lined with non-conductive material. Its energy content is approximately 1 J [91] and is therefore capable of igniting dust/air mixtures.

Propagating brush discharges can be prevented using conductive materials of construction.

Discharges from conical piles may, based on today's knowledge, occur from conveying highly chargeable, coarse particles at high velocities into vessels or silos. The discharge occurs at the surface of the highly charged conical pile and progresses towards conductive parts. If, in addition, fine dust is present, then there may be danger of an explosion. It can be assumed that this type of discharge has caused some explosion incidents in practice. From the products involved, e.g., sugar, wood, starch, corn starch, and epoxy dust, the energy level of a conical pile discharge can be estimated at <10 mJ [90]. An ignition has to be assumed possible for dust/air mixtures with a relatively low LMIE and for hybrid mixtures.

Conical pile discharges may be prevented through a reduction in feed rate and velocity. Preventive (inerting) or design measures also may have to be used.

It is possible to visualize a lightning-like discharge (Fig. 179) from a charged dust cloud towards a grounded object. Based on today's knowledge it is unlikely that such a discharge is possible in technical equipment. The sizes as well as the charges are too small in comparison with nature.

It has been experimentally proven [89] that lightning-like discharges are not to be expected in cylindrical vessels below 3-m in diameter even with unlimited height. In larger vessels there is no ignition danger if the overall field intensity is less than 500 kV/m [23].

5.2.6 Static Electricity

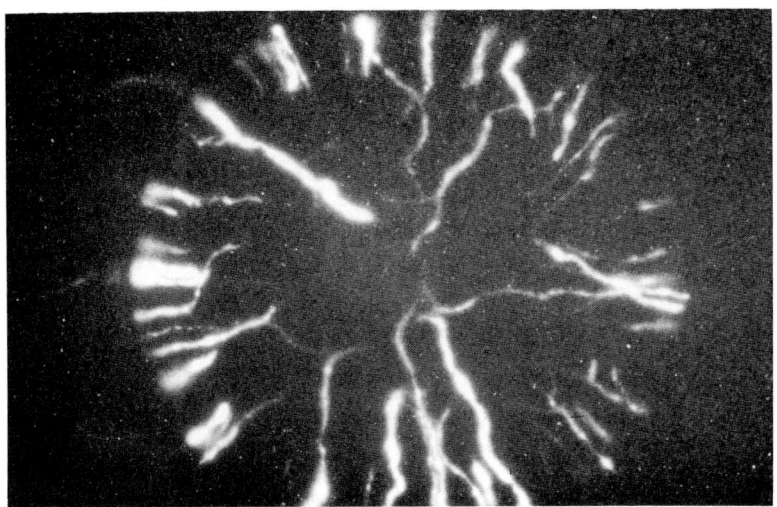

Fig. 178. Conical pile discharge

Fig. 179. Lightning-like discharge

5.2.7 Conclusions

There are a number of possibilities for applying preventive explosion protection to avoid dust explosions in practice. One possibility is limiting the dust concentration. If it is possible to maintain the dust concentration in the equipment outside the explosible range, then dust explosions are prevented.

In accordance with the guideline "explosion protection" [80], inerting is one of the measures which prevent the formation of dangerous, explosible atmospheres. This can be accomplished by admixing gaseous inert matter to the combustion air in the equipment to be protected. Inerting is the surest protective measure against the start of dust explosions. Therefore, it is a major tool for preventive explosion protection. In administering inerting by admixing nitrogen to the combustion air, it is important to know the limiting oxygen concentration (LOC) for industrial applications which will just prevent a dust explosion. Such a concentration is determined experimentally. Over a wide range it is independent of the initial pressure, the starting pressure of a dust explosion [93], but dependent upon the size of the test vessel (Fig. 180). The limits obtained in the standardized 1-m³ vessel are the guiding concentrations for practical applications; therefore, the values obtained in smaller apparatus have to be adjusted accordingly. By deducting a 2 vol% safety margin, the maximum allowable oxygen concentration is obtained which has to be maintained for inerting with nitrogen.

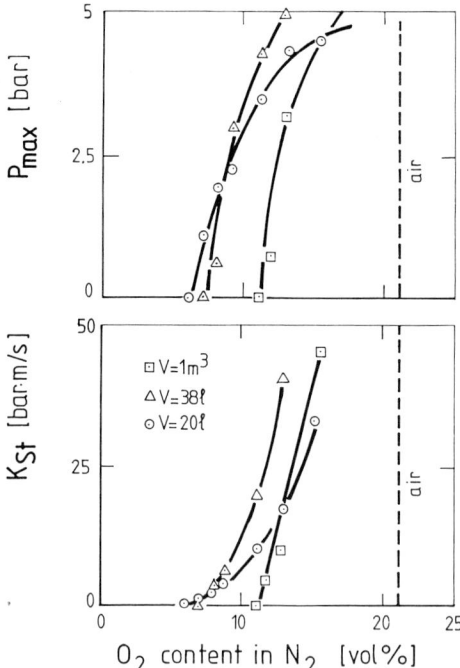

Fig. 180. Additives: influence of the size of the test vessel upon limiting oxygen concentration (E = 10,000 J)

Inerting can also be accomplished by using a vacuum or by admixing powdery inert dusts to the combustible dust. The use of the protective measure "prevention of effective ignition sources" has to be done with expert judgement in justified cases. It calls for expert judgement of the ignition capabilities of mechanically generated sparks, hot surfaces, glowing particle nests, and electrostatic discharges as the potential ignition sources. However, the knowledge accumulated from extensive research over the past years is now sufficient for reliable application of this safety measure in industrial practice.

5.3 Explosion Protection Through Design Measures

5.3.1 Preliminary Remarks

Explosion protection by design measures is always required whenever the goal of avoiding explosions through the application of preventive measures cannot be reached with an ample margin of safety. Such measures do not, as already stated, prevent the dust explosion, but instead reduce its effects to an acceptable level. It will be made certain that no personnel will suffer and that the protected equipment will generally be useable shortly after the explosion. Therefore, all equipment must be built "explosion resistant" in order to stand up to the anticipated explosion pressure. The relevant guidelines describe the protective design measures [23, 55, 80] based on the maximum explosion pressure obtained in a closed test vessel following an agreed-upon test method (see sect. 4.3.1.5). In practice, the actual pressure may be lower due to larger particle sizes (Fig. 74), deviation from the optimum dust concentration (Fig. 60), higher water content of the dust (Fig. 71), elevated temperature (Fig. 80), or only partial filling with explosible dust/air mixture of the equipment which has to be protected. It may also be exceeded because of a higher initial pressure (Fig. 79) or elevated oxygen concentration [60]. If the anticipated explosion pressure is lower than the maximum explosion pressure, then either pressure may be used for the design of the protective measure. However, if the anticipated explosion pressure is higher than the maximum explosion pressure, then the higher value has to be used [85]. The following protective design measures can be applied:

– pressure-resistant construction for the anticipated explosion pressure
– pressure-resistant construction for a reduced explosion pressure in conjunction with explosion pressure venting or explosion suppression.

The above-mentioned reduced explosion pressure is also referenced to the expected explosion pressure in the closed vessel.

The following remarks are basically contingent upon the maximum explosion pressure p_{max} determined in a closed vessel using an agreed-upon test method (see sect. 4.3.1.5). The vessels and the equipment which have to be protected can be designed in a pressure-resistant or pressure-shock-resistant fashion (Fig. 181).

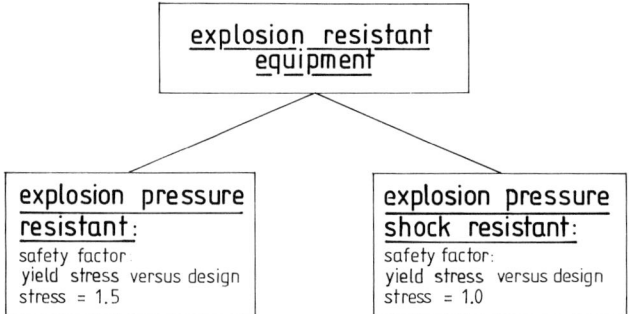

Fig. 181. Design possibilities for explosion resistant equipment using ductile materials of construction

The explosion pressure-resistant vessel withstands the anticipated pressure without deformation. The rules and regulations of the pressure vessel code apply to the design and to the manufacturing. The design pressure will either be the maximum explosion pressure p_{max} or the reduced maximum explosion pressure $p_{red,max}$ (see sect. 5.3.3).

The design of explosion pressure shock-resistant vessels is based on a concept dating back to 1971 for vessels and equipment which are rarely exposed to dust explosions. The construction principle combines that of pressure vessels (consisting of rotationally symmetrical parts) which have been used for 100 years in the chemical industry, and atmospheric vessels without pressure exposure capability.

The rules which apply for pressure vessels are used as much as possible for the pressure shock design. To a certain degree deviations are accepted in order to arrive at an economical construction which will meet the set goal of preventing the rupture of the vessel in case of an explosion. The major difference between a pressure-resistant and a pressure shock-resistant vessel is that the latter utilizes the strength of the material of construction to a higher degree [96, 97]. In such a case, the safety factor for ferritic steel is 1.0 instead of 1.5 and in case of an explosion, the material is stressed up to its yield strength. The same applies for austenitic steels, i.e., in case of an explosion, the material will be stressed up to a 2% strain. The following correlation exists for the explosion pressure-resistance versus the explosion pressure shock-resistance:

Explosion pressure shock-resistance = 1.5 explosion pressure-resistance

In the case of a vessel buit for a design pressure of 6 bar, its explosion pressure shock-resistance will be 9 bar. Such a vessel in a closed state will withstand explosions of practically all organic dusts (Fig. 88).

If such a vessel is available having a code design pressure of 2 bar, it may be used in conjunction with the measure explosion pressure venting for a maximum reduced explosion pressure $p_{red,max} = 3$ bar. In exceptional cases, permanent deformation has to be expected for pressure shock-resistant vessels, when subjected to explosions. There is still a substantial safety margin up to the ultimate tensile strength which amounts to 60–90% depending upon the steel [96, 97]. The sole application of the above-mentioned design measure is generally not sufficient in itself in order to meet the explosion danger. For equipment systems, it is often necessary to prevent the propagation of an explosion from equipment which is protected through design measures into operating areas or equipment which are protected by preventive measures and, therefore, not pressure-rated. Such an approach is referred to as the "technical disengagement of systems subject to explosions".

The application of protective design measures calls for an automatic shut-off of the product flow and of the equipment exhaust. It should be stressed that an absolutely safe installation does not and will not exist [98, 99] despite all the design and other measures which are used as state-of-the-art. There exists always a residual risk which is hard to quantify. Whether or not it is justifiable depends mainly upon the experience and the results from systematic and experimental tests. It is mainly influenced by:

- the intentionally assumed risk, e.g., erroneous behavior of people in operating and supervisory functions
- the technical failure of the selected protective measure
- undetected dangers during the safety analysis based on today's knowledge.

Such residual risks cannot be reduced indefinitely nor compensated for through further substantial, additional measures. On the other hand, productivity and economy should not suffer excessively because of the safety measure considered necessary. In the case where the residual risk is considered unacceptably high, the process may have to be abandoned. Therefore, it is the daily duty of the safety representative to address the question about the right approach.

5.3.2 Explosion Pressure-resistant Design for the Maximum Explosion Pressure

5.3.2.1 Explosion Pressure-resistant Design

One possibility for protecting equipment and apparatus from the effects of pressure from dust explosions is a design for the maximum explosion pressure. Such equipment is mainly built from rotationally symmetrical parts using all code calculations and specifications for pressure vessels [100], including inspection and stamping.

The operating pressure in the dust-handling vessels and equipment is generally between +40 and −20 mbar. The maximum explosion pressure is therefore not equivalent to the allowed operating pressure nor to the expected overpressure. In cases where the operating pressure is equivalent to the explosion pressure, the expected explosion pressure would be 8–10 times higher than the operating pressure (Fig. 79).

Explosion pressure-resistant construction requires a test, e.g., with water as per the requirements of the "AD-Merkblätter" [100].

164   5 Protective Measures Against the Occurrence and Effects of Dust Explosions

Despite the use of explosion pressure-resistant equipment for the full explosion pressure or other design measures outlined later, everything possible must be done to prevent effective ignition sources because loss of product and interruption of production are equally undesirable. The entrance of tramp metal into milling and size reduction equipment can be prevented to a large degree, e.g., through the use of precrushers or safety crushers with slip clutches or inductive metal detectors.

5.3.2.2 Explosion Pressure Shock-resistant Design

Vessels and equipment can be protected from the pressure resulting from dust explosions through an explosion pressure shock-resistant construction. But this should only be done if an explosion is expected rarely.

Again, the rules valid for pressure vessels are applied as much as possible for said vessels and apparatus, with some deviation in a few points. Such deviations include the use of flat walls and the utilization of a higher value for the design strength of the given material of construction (see sect. 5.3.1). The actual operating pressure is relatively low for the vessels or apparatus in which an explosible dust/air mixture may exist (sect. 5.3.2.1). Therefore, they are not covered by the pressure vessel code.

Figures 182 and 183 depict equipment which is built according to an explosion pressure shock-resistant design.

Fig. 182. Mill housing built to be explosion pressure shock-resistant [96]

5.3.2 Explosion Pressure-resistant Design for the Maximum Explosion Pressure

Fig. 183. Double rotary air lock built to be explosion pressure shock-resistant [96]

Generally, pressure vessels are subjected to a pressure test with water. For ductile materials, the test pressure is 0.9 times the design pressure, and for cast iron twice the design pressure. An explosion test can also be made in the form of a specific prototype test.

Figure 184 (left) shows a small container for bulk material with flat surfaces and Fig. 184 (right) the same container after exposure to an aluminum dust explosion with a maximum explosion pressure $p_{max} = 10$ bar. The walls are deformed, but the container did not rupture. It can therefore be classified as explosion pressure shock-resistant up to the test pressure and be used, e.g., as a receptacle for dust explosible products from milling systems which are protected through appropriate design measures. A number of dust explosion incidents in practice have documented that the tested container has met all the required demands. Some guidelines which have to be watched in the design of explosion pressure shock-resistant vessels are given in [102] and [103].

166 5 Protective Measures Against the Occurrence and Effects of Dust Explosions

Fig. 184 a/b. Small container for bulk material (V = 1-m³) (**a**) as received; **b**) after explosion test)

5.3.3 Explosion Pressure-resistant Design for a Reduced Maximum Explosion Pressure in Conjunction with Explosion Pressure Venting

5.3.3.1 Preliminary Remarks

The term explosion pressure venting covers, in a broad sense, all measures which allow initially closed vessels to open temporarily or permanently in a safe direction [55] at the start or after a certain propagation of a dust explosion. Therefore, the question is how to control most efficiently the results of such explosions in vessels through pressure-venting systems. (Even when pressure-venting systems are available, everything possible should be done to prevent effective ignition sources). The actuation of the pressure-relief system may cause air pollution for a limited time and area. Therefore, such a measure is not suited for vessels containing poisonous or corrosive chemicals.

Pressure venting, through the timely opening of a defined area, must prevent unacceptably high explosion pressures in a vessel, silo, or pipeline. The maximum explosion pressure will be reduced below the design pressure of the equipment by venting unburnt mixtures and gases of combustion into the atmosphere.

5.3.3 Explosion Pressure-resistant Design with Explosion Pressure Venting

Pressure-venting systems can be built either for only one exposure, e.g., rupture discs, or multiple exposures, e.g., explosion flaps (doors).

The safety measure "explosion pressure venting" presumes a certain pressure rating of the vessel. The vessels which are to be protected must be designed as "explosion pressure-resistant" or "explosion pressure shock-resistant" as per sect. 5.2.4.1. The design pressure is a reduced maximum explosion pressure $p_{red,max}$ compared with the maximum explosion pressure p_{max}. The pressure reduction is accomplished through a properly sized vent system. Attention has to be paid to ensure that all parts which are exposed to the explosion pressure be included in the considerations for the suitable protective measures.

5.3.3.2 Explosion Pressure Venting of Vessels

The reduced explosion pressure p_{red} and the corresponding pressure rise (dp/dt) Fig. 185 are the measures of the violence of a dust explosion in vented vessels at any dust concentration. The maximum values of explosion parameters are determined through tests with the combustible dust over a wide range of concentrations (Fig. 186). These maximum values are:

- reduced maximum explosion pressure $p_{red,max}$ and
- reduced maximum rate of pressure rise $(dp/dt)_{red,max}$

They may occur at the same concentrations which gave the maximum explosion characteristics in the closed vessel, or the concentrations may differ.

In contrast to the closed vessel (Figs. 59/60), the course of an explosion in a pressure-vented vessel is described by the reduced maximum pressure and reduced maximum rate of pressure rise instead of the maximum pressure and maximum rate of pressure rise.

Shortly after the development of a new dust-testing procedure (Fig. 23) which reflects practical conditions for the turbulence of dust/air mixtures at the time of ignition, testing was begun to answer the questions pertaining to pressure venting of combustible dust in vessels $V = 1-60\text{-m}^3$ (Figs. 187/188) [104].

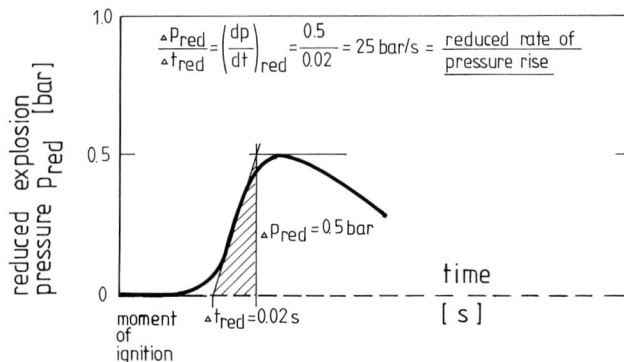

Fig. 185. Definition of the reduced rate of pressure rise $(dp/dt)_{red}$ of a pressure-vented dust explosion

5 Protective Measures Against the Occurrence and Effects of Dust Explosions

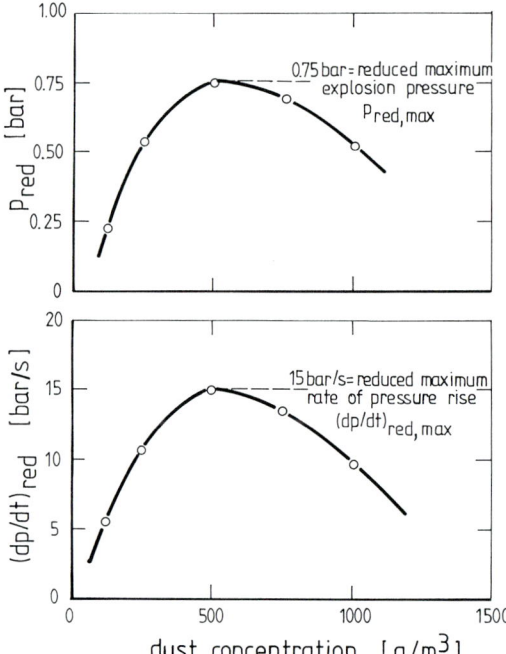

Fig. 186. Determination of the optimum values of the reduced maximum explosion characteristics of a combustible dust with pressure venting

Fig. 187. Dust explosion in a pressure-vented 1-m³ vessel

5.3.3 Explosion Pressure-resistant Design with Explosion Pressure Venting

Fig. 188. Dust explosion in a pressure-vented 60-m³ chamber

The knowledge gained from those dust explosion tests was the basis for the VDI guideline 3673 "Pressure release of dust explosions" (Fig. 189) [55]. The guideline includes nomograms (Fig. 190) which give in a convenient way the required relief area for known vessel volume, vessel design pressure (equivalent to the reduced maximum explosion pressure $p_{red,max}$), and the dust explosion class St for static venting pressures $p_{stat} = 0.1$–0.5 bar.

The nomograms are valid:

– for the case of pressure venting with a safety membrane (Fig. 191) or a metallic rupture panel (Fig. 192)
– for dust having a maximum explosion pressure $p_{max} = 10$ bar for dust explosion classes St 1 and St 2, as well as $p_{max} = 12$ bar for the dust explosion class St 3

In order to determine the vent areas for any K_{St}-value, a mathematical adjustment of the measured values has been made. The resulting nomograms can also be taken from the referenced VDI guideline. There are differences in the resulting vent sizes between the nomograms which can be justified from a safety point of view. Extensive tests have shown that the deviation of the test data markedly increases with increased maximim reduced explosion pressure. Therefore, $p_{red,max}$ was limited in the nomograms to $p_{red,max} = 2$ bar gage. Upon actuation of the relief device, a substantial flame formation and pressure build-up has to be expected, even for relatively small vessels (Fig. 187), due to the exhaust of combustible dust. The vented, at first unburnt dust will then be ignited outside the vent opening. Therefore, the flame propagation will be larger the smaller the static venting pressure of the relief device. If pressure venting is applied to vessels or equipment inside a building, it is imperative that the pressure be vented through ductwork outside in a safe direction in order to protect personnel and production areas

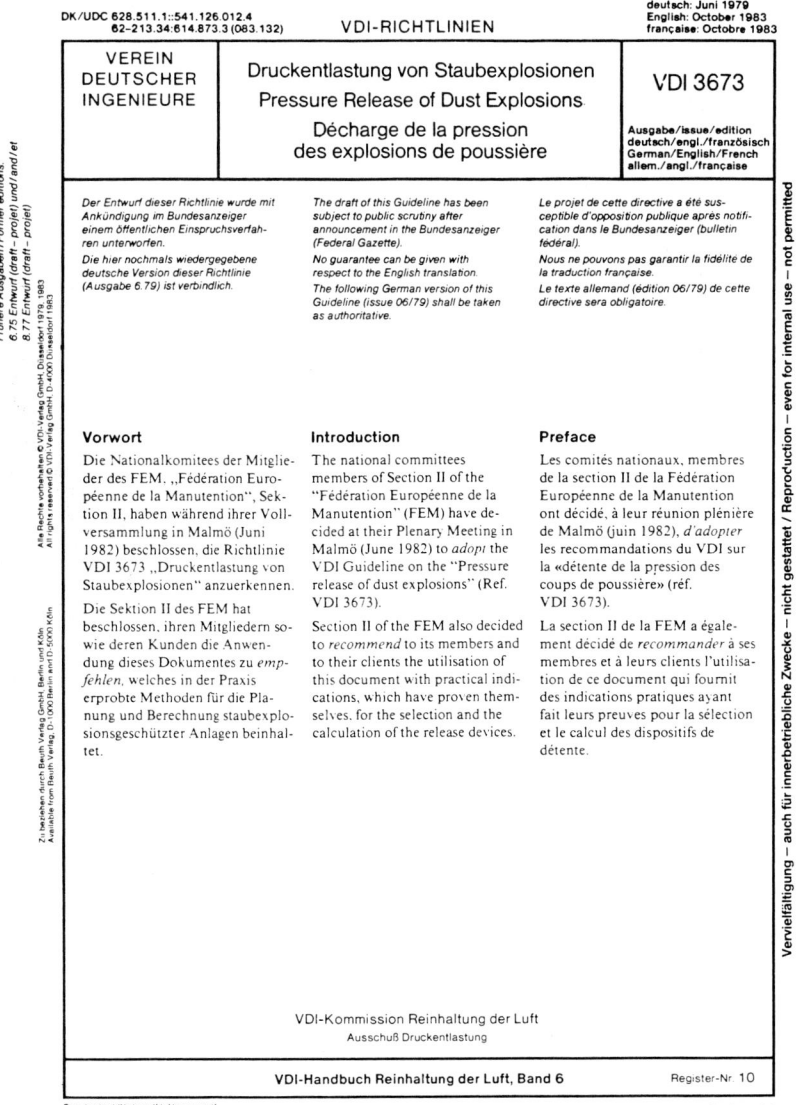

Fig. 189. Cover sheet VDI guideline 3673 "Pressure release of dust explosions"

5.3.3 Explosion Pressure-resistant Design with Explosion Pressure Venting

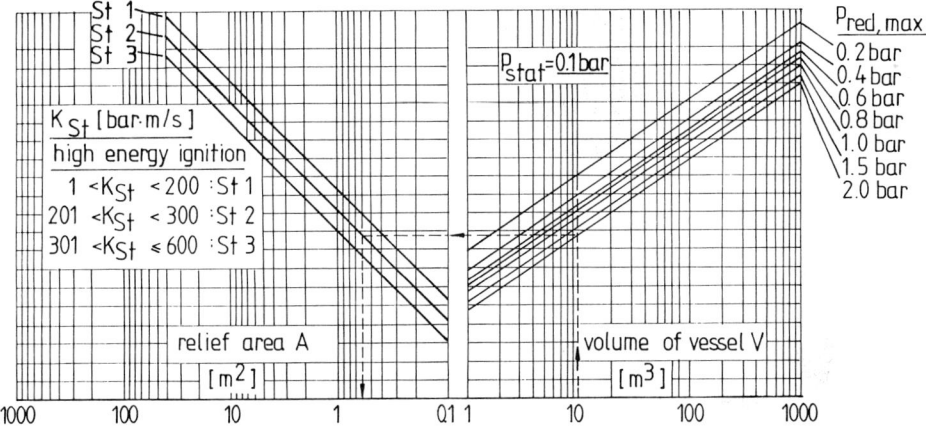

Fig. 190. Nomogram for the determination of the pressure relief area for dust explosions in vessels (Example: static venting pressure $p_{stat} = 0.1$ bar)

Fig. 191. Safety membrane made out of plastic foil (left: before actuation; right: after actuation)

172   5 Protective Measures Against the Occurrence and Effects of Dust Explosions

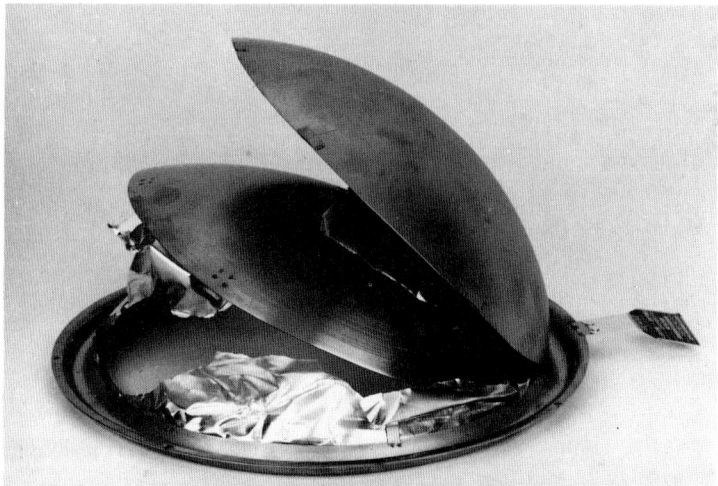

Fig. 192. Three-part rupture panel, Rembe (Brilon FRG)

(Fig. 193). However, such ductwork will substantially increase the violence of the explosion and the pressure in the protected vessel. This amplification of the course of the explosion is caused by the influence of the secondary explosion in the vent duct upon the explosion in the vessel proper.

Fig. 193. Vent ducts attached to the vent openings of filter housings

5.3.3 Explosion Pressure-resistant Design with Explosion Pressure Venting 173

Upon actuation of the vent, the initially dust-free vent pipe will be filled with unburnt dust ahead of the flame front. The explosible mixture will then be ignited by the trailing flame jet from the protected vessel. The pressure build-up of the secondary explosion restricts the exhaust and markedly increases the reduced maximum explosion indices. The increase is especially pronounced if the flame velocity in the vent duct reaches or exceeds sonic velocity.

This is always to be expected if the length of the vent duct is more than 3 m. The reduced maximum explosion pressure with unobstructed venting determines the increase of the pressure in the protected vessel with a vent pipe. The corresponding values can be taken from Fig. 194.

Tests over the past years with various combustible dusts [105, 106] have confirmed that the correlation shown in Fig. 194 is not affected by the mounting of the ducts to the vents of such equipment as sifters, vessels, filter housings, or mills. Because of the anticipated increase of the maximum reduced explosion pressure in the vented vessel, either the vent area or the explosion resistance (pressure shock resistance or explosion pressure resistance) have to be increased. It has to be at least 0.5–1.0 bar for large vent areas and pressure build-ups in the range of the static vent pressure. Vessels with a lower pressure rating cannot be practically protected using explosion pressure venting through vent ducts. Vent ducts have to be as short as possible and must lead in the direction of the discharge directly into an open area.

Bent pipes result in an unforeseeable additional increase of the reduced maximum explosion pressure. In exceptional cases, bent pipes are acceptable if there is a straight run for quite a distance immediately after the vent opening. The vent pipes must have at least the same cross section as the vent area.

A round cross section is preferred. The explosion pressure shock resistance (or explosion pressure resistance) of the pipe has to be equivalent to that of the vessel

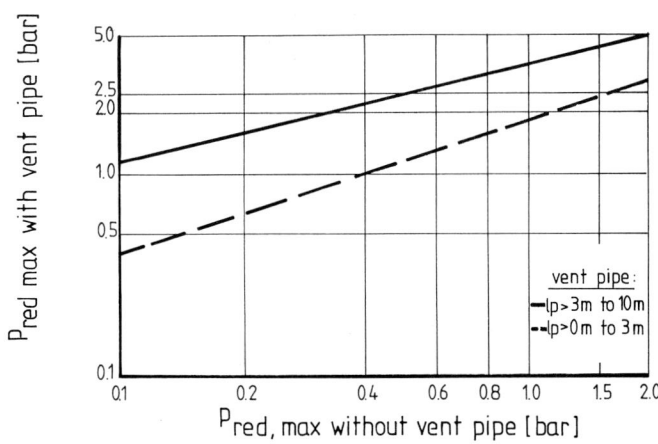

Fig. 194. Influence of vent pipes upon the reduced maximum explosion pressure of combustible dusts in vented vessels

to be protected. If there are inspection openings in the vicinity of the vent area, they, as well as the closures, require the same design strength.

In designing vent pipes, one has to consider that dust explosions in pipelines may be similar to detonations (Fig. 138). Pressures of 20–30 bar will then be reached (Fig. 139). Therefore, it is sensible to limit the length of the vent pipes to approximately 6–10-m. Otherwise extremely high pressures are to be expected at the mouth of the vent pipe. The pressure rating of a pipe has to be at least 10 bar if such a limitation cannot be maintained due to urgent circumstances. In case rupture panels are substituted with explosion flaps (doors) (Fig. 195) for the pressure venting of the vessels, efficient venting may be limited by the mass of the flap, depending upon the violence of the dust explosion. Therefore, explosion flaps have to be built as light as possible and be subjected to a performance test before being used in practical applications. At the same time, the venting capability must be compared with that of a rupture panel (Fig. 196). Any limiting influence can be met by increasing the pressure rating of the vessel or the size of the vent area.

An increase in the static venting pressure may be caused by corrosion, improper painting of movable parts, icing, or snow loads. The venting device has to be checked in predetermined intervals with respect to functional condition and mobility. Explosion pressure venting is used very frequently as a safety measure. The

Fig. 195. Explosion flap (door) (A = 0.5-m^2) on top of the vent opening of a 10-m^3 vessel

5.3.3 Explosion Pressure-resistant Design with Explosion Pressure Venting

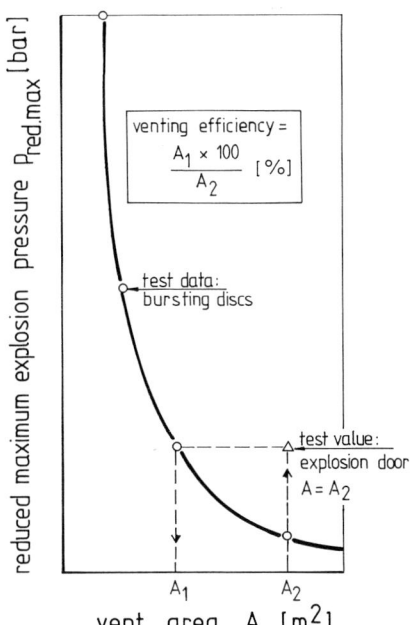

Fig. 196. Determination of the venting capability of an explosion flap (door) for dust explosions

pressure venting process not only consists of pressure build-up and flame propagation outside the vessel but also creates a recoil force (Newton's force of reaction principle). Such a force may pull vessels or equipment off their anchors, tip them, and cause substantial secondary damage, even if the vessel is properly designed for the given reduced maximum explosion pressure. VDI guideline 3673 "Pressure release of dust explosions" points towards the dangers which may be caused by recoil forces.

Several publications deal specifically with the stated problem [107–110]. Ritter uses a thermodynamic method for the calculation of the recoil force which makes allowances for the vent flow [108]. Experimental results have led to the following equation, which is based on a static venting pressure $p_{stat} = 0.1$ bar for a vent system consisting of a safety membrane or rupture panel.

$$F_R = \alpha \cdot 100 \cdot A \, (p_{red,max} - p_{atm})$$

F_R = recoil force [kN]
α = dynamic coefficient
A = vent area [m^2]
$p_{red,max}$ = reduced maximum explosion pressure [bar absolute]
p_{atm} = atmospheric pressure [bar absolute]

The value of the dynamic coefficient fluctuates in line with the test arrangement.
For a rigid support of vented vessels having sizes $V = 0.25$–25-m^3, the range of $\alpha = 1.19$–1.21. For a flexible support, as it may exist for vessels mounted to the ceiling of a manufacturing building, the dynamic coefficient for a 10-m^3 vessel was de-

termined at $\alpha = 1.13$. The experimentally determined value of the recoil force per area is therefore in the middle between the calculated values using the thermodynamic approach and the stationary approach, which ignores the vent flow (Fig. 197). This is not valid for vent pipes which are downstream of vent openings. In such a case, the values from the stationary method will be approached. Such a phenomenon may be explained by the flow restrictions given by the vented vessel.

The recoil forces are independent of the kind of combustible dust; however, they are influenced by the dust-specific K_{St}-value, which reflects the pressure behavior per time the explosion takes in the vented vessel.

For a given dust and a defined vessel size, the reduced maximum explosion pressure decreases hyperbolically with increased vent area. Recoil forces, however, behave differently. They first increase and reach an "optimum recoil force" for an "optimum vent area" and then decrease with increasing vent areas. For several dusts, it was found that the corresponding reduced maximum explosion pressure amounted to $p_{red,max} = 3.3$ bar.

For very large vent areas and direct vessel venting, the reduced maximum explosion pressure approaches more and more the static venting pressure. Therefore, the recoil force will reach a minimum and thereafter increase proportionally with the vent area. The so-called "optimum recoil force" will not be reached with designs based upon the VDI guideline 3673 "pressure release of dust explosions" and with no vent lines since the nomograms are restricted to a reduced maximum explosion pressure $p_{red,max} = 2$ bar because of already-stated reasons. With existing vent lines, however, depending upon their length, the optimum recoil force may be

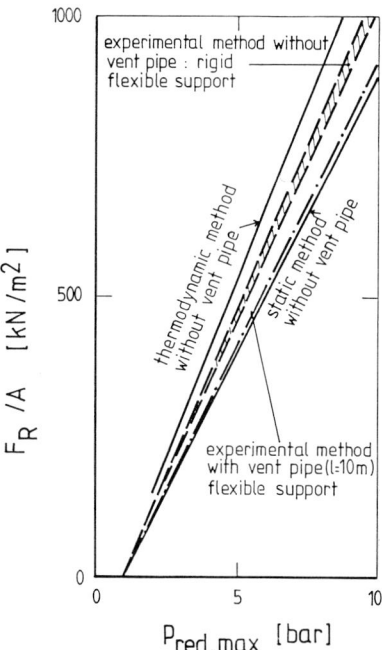

Fig. 197. Measured and calculated recoil forces per area for pressure-vented dust explosions

5.3.3 Explosion Pressure-resistant Design with Explosion Pressure Venting

Fig. 198. Spray drier with explosion pressure venting

reached or exceeded for oversized vent areas. For more details pertaining to recoil forces, consult the listed publications. Figure 198 shows a spray drier protected with explosion pressure venting. Fluidized bed driers are also often protected with the design safety measure "explosion pressure venting". In accordance with outline 5.1, they have to be explosion-resistant for a certain reduced maximum explosion pressure. As the first basic explosion tests were made more than 10 years ago, it was necessary to confirm the earlier results [111, 112]. In particular, the influence of the shape of the vent pipes upon the course of dust- and hybrid explosions was studied in a certain fluidized bed drier to be protected.

The test results coincide with the requirements of the safety rules issued by the "Berufsgenossenschaft" [113] with regard to size of the vent area and explosion resistance of the apparatus subjected to dust explosions. In addition, it was found that:

- the product filter (filter bag) will only amplify the explosion if there is no vent pipe (Figs. 203 and 204)
- a vent pipe with enlarged cross section will not improve the venting process
- the reduced explosion pressure in the fluidized bed drier linearly increases with the length of the vent pipe and in the absence of a product filter will follow the VDI guideline 3673 "pressure release of dust explosions".

178 5 Protective Measures Against the Occurrence and Effects of Dust Explosions

Fig. 199. Fluidized bed drier with vent pipe without change in cross section

Fig. 200. Fluidized bed drier with vent pipe having enlarged cross section

– a reduced cross section (Fig. 201) markedly elevates the reduced maximum explosion pressure to a much greater degree than a bend in the vent line (Fig. 202).

In conjunction with hybrid explosions, it was found [112] that the area requirements called for by the "Berufsgenossenschaftliche" safety rules [113] are not suffi-

5.3.3 Explosion Pressure-resistant Design with Explosion Pressure Venting

Fig. 201. Fluidized bed drier with vent pipe having reduced cross section

Fig. 202. Fluidized bed drier with vent pipe with bend

cient. This can only be alleviated by increasing the vent size and, if not possible, by increasing the explosion resistance of the apparatus to be protected.

Mechanical mills are always to be considered as possible ignition sources for explosible dust/air mixtures as per today's knowledge. This is also true for non-inerted coal pulverizers, in which an ignition of the coal dust mixtures has to be ex-

Fig. 203. Dust explosion in a fluidized bed drier without filter bag and vent pipe

Fig. 204. Dust explosion in a fluidized bed drier with filter bag and without vent pipe

pected. The results of dust explosion tests [114] in an initially unprotected pulverizer with volume $V = 8$-m^3 have shown that the system will survive the explosion without damage, provided coal dust has a maximum explosion pressure $p_{max} \leq 8.1$ bar and a dust-specific characteristic $K_{St} \leq 105$ bar·m/s. The large inserts and the venting through three pipes must restrain the explosion from full development.

5.3.3 Explosion Pressure-resistant Design with Explosion Pressure Venting

Low values have been recorded in the pulverizer proper as well as in the feeder [$p_{red,max} \leq 1.2$ bar; $(dp/dt)_{red,max} \leq 16$ bar/s]. Substantial flame propagation and escaping combustion gases mixed with unburnt dust/air mixtures were observed at the mouth of the coal dust pipe.

An explosion with a more violently reacting coal dust ($p_{max} = 9.4$ bar; $K_{St} = 185$ bar·m/s), however, resulted in a pressure of $p_{red,max} = 2.4$ bar in the housing, deforming the inspection doors, and a pressure of $p_{red,max} = 2.5$ bar in the feeder, leading to its destruction (Fig. 205).

In order to remedy the situation, a suitable vent opening was installed on top of the feeder (Fig. 206) which was closed with a safety membrane [115]. Such modification not only markedly reduced the explosion characteristics in the pulverizer with the less violently reacting coal dust (Fig. 207), but it also withstood the explosion of the violently reacting coal dust, even with a vent pipe attached to the vent opening.

The above statements may certainly not be generalized. They show, however, that cost-efficient design solutions may be found from properly conducted tests which improve certain unprotected systems.

Fig. 205. Coal pulverizer: Feeder ripped on one side after dust explosion

Fig. 206. Coal pulverizer: Arrangement of the vent system for the feeder

The development of the VDI guideline 3673 "pressure release of dust explosions" (Fig. 190) was based upon the assumption that the cubic law was not only valid for closed apparatus (Fig. 64) but also for vented apparatus.

$$\left(\frac{dp}{dt}\right)_{red, max. V_1} \cdot V_1^{1/3} = \left(\frac{dp}{dt}\right)_{red, max. V_2} \cdot V_2^{1/3}$$

and

$$A_2 = \left(\frac{V_2}{V_1}\right)^{2/3} \cdot A_1$$

5.3.3 Explosion Pressure-resistant Design with Explosion Pressure Venting

Fig. 207. Coal pulverizer with pressure-vented feeder: coal dust explosion

In other words:

- the product of reduced, maximum rate of pressure rise times the cube root of volume is constant and
- for a constant reduced maximum explosion pressure referenced to a given static venting pressure, the vent area A_2 for a volume V_2 can be calculated from the known area A_1 of volume V_1.

The results from explosion pressure venting tests from vessels $V = 2.4–250\text{-m}^3$ (Fig. 208) with combustible dusts confirmed this only conditionally [104].
It was shown that with increased vessel size:

- for dusts belonging to the dust explosion class St 1 ($K_{St} \leq 200$ bar·m/s), the area requirements increase more rapidly and the reduced maximum rate of pressure rise decreases more slowly than projected with the cubic law.
- the same statement is true for dusts belonging to the dust explosion class St 2 ($K_{St} \leq 300$ bar·m/s) for vessel sizes $V < 25\text{-m}^3$. For larger vessel sizes, however, the area requirement and the explosion violence follow the cubic law.

Fig. 208. Dust explosion in a vented 250-m^3 vessel

As a consequence (Fig. 209), the area requirements of very large vessels (V>600 m^3) at a given reduced maximum explosion pressure are independent of the dust-specific characteristic K_{St} and probably follow the cubic law.

This phenomenon may possibly be explained as follows: The amount of initially unburnt dust which is pushed out of the vent opening, ahead of the flame front, is much larger for dusts belonging to the dust explosion class St 1 than the one for dusts belonging to class St 2. This is documented by film sequences showing venting action. Therefore, the secondary explosion initiated in the free atmosphere by the departing flame is much more severe for St 1 dusts than for St 2 dusts. Pressure measurements outside the vessel point also in that direction. Therefore, it is conceivable that the pressure produced by the dust's secondary explosion will obstruct the venting process and raise the level of the reduced maximum explosion pressure.

This is more pronounced for less violently reacting dusts of the dust explosion class St 1 than for the more violently reacting St 2 dusts and may therefore explain the behavior of the area requirement in Fig. 209. These statements are based on a maximum explosion pressure p_{max} = 9 bar, whereas the nomograms of the VDI guideline 3673 [55] allow p_{max} = 10 bar. A comparison of the area requirements is possible in Figs. 210 and 211 for two constant values for the reduced maximum explosion pressure.

5.3.3 Explosion Pressure-resistant Design with Explosion Pressure Venting

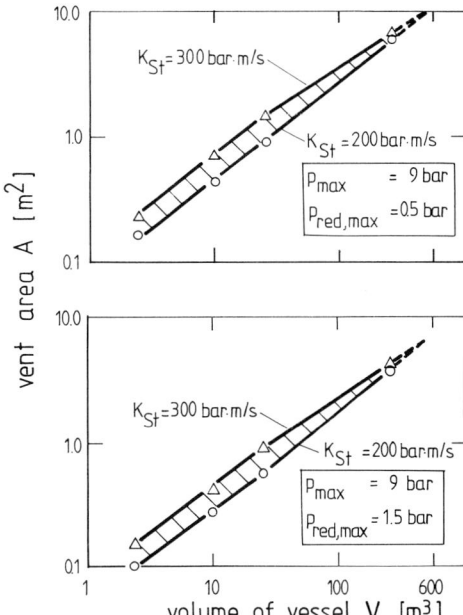

Fig. 209. Comparison of area requirements depending on vessel size for St 1 and St 2 dusts ($p_{stat} = 0.1$ bar)

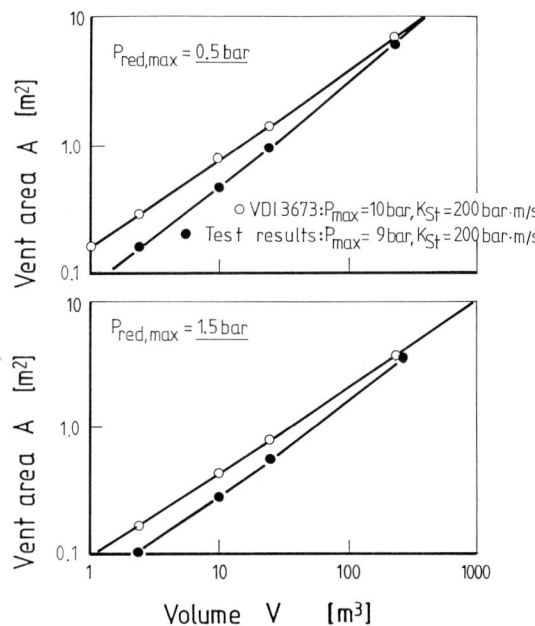

Fig. 210. Comparison of area requirements for VDI-guideline 3673 with newer test results ($p_{stat} = 0.1$ bar; St 1)

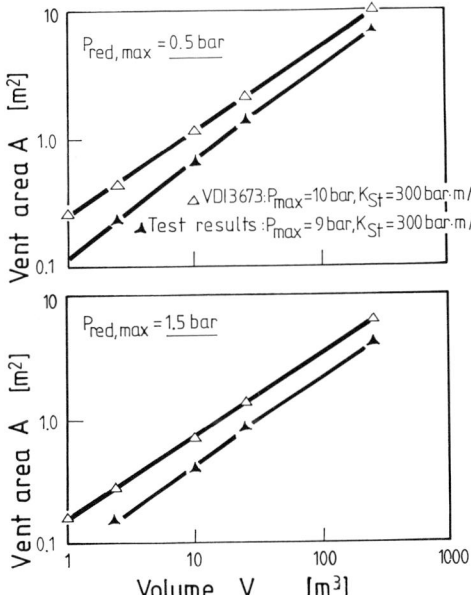

Fig. 211. Comparison of area requirements for VDI-guideline 3673 with newer test results ($p_{stat} = 0.1$ bar; St 2)

The reduction in area requirements, especially for smaller vessels, in comparison to the VDI guideline is most pronounced for dusts belonging to the dust explosion class St 1 (Fig. 210) which includes a majority of the dusts tested so far (Fig. 87). The area reduction decreases with increased vessel volume and lies between 41% and 5%. For dusts belonging to the dust explosion class St 2 (Fig. 211), an area reduction is possible for the whole range of volumes. The reduction decreases slightly from 50% to 36%.

In order to transfer the results from the newer explosion pressure venting test into practice, all vent areas for a constant reduced explosion pressure were calculated for volume $V_1 = 1$-m³ and tabulated (Table 21). For any given vessel size V_2, shown in column 3, the required vent area A_2 can be calculated as follows:

$$A_2 = (V_2)^{2/3} \cdot A_1$$

It must be emphasized again that the equation is only valid together with Table 21 if the maximum explosion pressure of the dusts does not exceed $p_{max} = 9$ bar and the static venting pressure amounts to $p_{stat} = 0.1$ bar. If the maximum explosion pressure of a dust is more than $p_{max} = 9$ bar, then the calculation of the necessary vent area has to be in accordance with the nomograms of the VDI guideline 3673 "Pressure release of dust explosions" [55] (nomograms, see Addendum 8.1.1).

5.3.3 Explosion Pressure-resistant Design with Explosion Pressure Venting

Table 21. Specific vent area, referenced to $V_1 = 1\text{-m}^3$, as the basis for the calculation of vent areas for vessels exposed to dust explosions. (Maximum explosion pressure $p_{max} = 9$ bar, static venting pressure of safety membrane $p_{stat} = 0.1$ bar)

p_{max} [bar]	K_{St} [bar·m/s]	Vessel volume [V] to be protected [m³]	$p_{red,max}$ [bar] 2.0	1.5	1.0 A_1 [m²]	0.5	0.25
9.0	200	$1 \leq V < 6$	0.048	0.056	0.068	0.089	0.120
		$6 \leq V < 18$	0.048	0.058	0.072	0.096	0.150
		$18 \leq V < 100$	0.058	0.063	0.075	0.110	0.160
		$100 \leq V < 600$	0.081	0.089	0.110	0.140	0.180
		≥ 600	0.081	0.089	0.110	0.160	0.240
9.0	300	$1 \leq V < 6$	0.068	0.079	0.096	0.127	>0.17
		$6 \leq V < 18$	0.068	0.079	0.096	0.142	>0.20
		≥ 18	0.084	0.098	0.120	0.170	0.240

5.3.3.3 Explosion Pressure Venting of Elongated Vessels (Silos)

Elongated vessels (silos) are defined as vessels having a height to diameter ratio larger than 5:1. They are subjected to a very pronounced axial flow, similar to pipelines (see sect. 4.5). There exist major differences from the course of an explosion in a closed cubic vessel, which have to be considered when sizing the vent area.

Here, it is necessary to allocate the total available roof area (cover area) for pressure venting (Fig. 212). Side vents or partial venting of the cover may result in the tearing off of the complete vessel cover (silo top) (Fig. 213).

Generally, in sizing the vent areas of oblong vessels (silos), the nomograms of the VDI guideline 3673 "Pressure release of dust explosions" [55] may be used (Fig. 190). However, the following restrictions have to be observed:

- the vent areas given by the nomograms may not be reduced, i.e., the cross section of the vessel (silo) determines the maximum volume which can be protected;
- said nomograms are only valid for volumes up to 1000-m³.

From these restrictions there arises a certain correlation between the diameter of the oblong vessel (silo) and its maximum height (considering the explosion resistance and explosion characteristics).

The explosion pressure-resistance (or explosion pressure shock resistance) especially has a decisive influence upon the allowable height. As the cross section

188   5 Protective Measures Against the Occurrence and Effects of Dust Explosions

Fig. 212. Silo top designed for pressure venting

Fig. 213. Partially vented silo top torn off after dust explosion

5.3.3 Explosion Pressure-resistant Design with Explosion Pressure Venting

(roof) of an oblong vessel (silo) represents the largest possible vent area A_2, the maximum volume V_2 which can be protected with such an area can be calculated from the cubic law:

$$V_2 = \left(\frac{A_2}{A_1}\right)^{3/2} \cdot V_1$$

The area requirement A_1 for the volume $V = 1$-m^3 can be taken from the nomogram. The reduced maximum explosion pressure $p_{red,max}$ and the static venting pressure p_{stat} of the venting device coincide for both vessels (V_1 and V_2); they are given and therefore constant. The consequences from the equation for the maximum allowed silo height for dust explosions are shown in Fig. 214 for two silo diameters (D = 4 m and D = 6-m). For a height $H \leq 5\,D$ (e.g., D = 4-m, $H \leq 20$-m; D = 6-m, $H \leq 30$ m), the nomograms are applicable without restrictions. For a height $H > 5\,D$ (e.g., D = 4-m, $H > 20$-m; D = 6-m, $H > 30$-m), the whole roof area of the silo has to become the vent area. The maximum height and with it the maximum volume of the silo to be protected will increase with increased pressure resistance of the vessel. For dusts belonging to the dust explosion class St 1 and for a static venting pressure $p_{stat} = 0.1$ bar, the height of the silo is only limited in case the volume of the silo exceeds 1000-m^3 (e.g., D = 4-m, $H \geq 80$-m; D 6-m, $H \geq 35$-m).

The recommendations of the VDI guideline 3673 "Pressure release of dust explosions" are based on knowledge pertaining to the course of an explosion of combustible dusts in pipelines. Other basic experimental facts were unavailable at the time the guideline appeared. With 20% of the incidents, silos and bunkers repre-

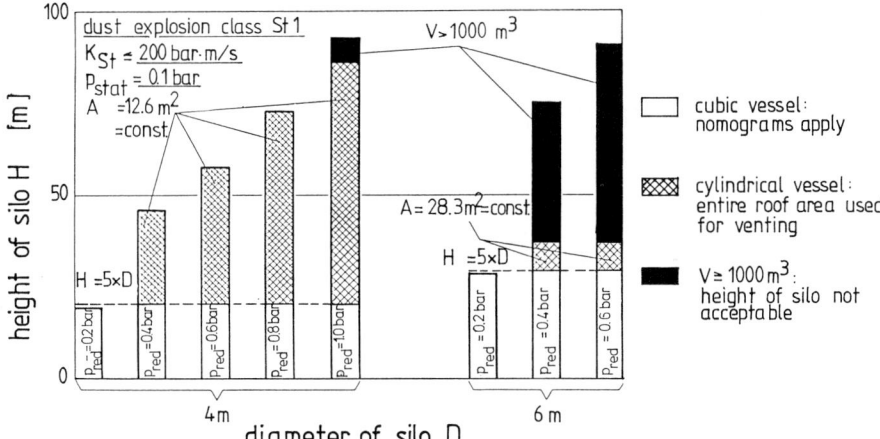

Fig. 214. Influence of the explosion resistance of silos upon the silo height when using nomograms

Fig. 215. Horizontally arranged steel silo (V = 20-m^3; H/D = 6.25)

Fig. 216. Vertically arranged reinforced concrete silo (V = 20-m^3; H/D = 6.25)

5.3.3 Explosion Pressure-resistant Design with Explosion Pressure Venting

sent the largest group involved in explosions (Figs. 6 and 7). Therefore, in 1981, dust explosion tests were started in vented silos $V = 20$-m³ (height/diameter ratio 6.25) in order to obtain information about the actual course of an explosion in a horizontal (Fig. 215) or vertical arrangement (Fig. 216) [116–119]. The mixture was tested in accordance with the agreed-upon procedure [55], utilizing rapid dispersion of the combustible dust out of a dust storage chamber plus initiation of ignition after a predetermined delay time with pyrotechnic igniters having an energy content $E = 10{,}000$ J.

While the reduced maximum explosion pressure decreases linearly in a vented cubic vessel with decreasing filling ratio (Fig. 217), such a decrease is noticeable only for a filling ratio <50% in the horizontal silo. This is caused by the unburnt mixture which is pushed ahead of the flame front. This is also the cause for the higher explosion violence in a silo compared with a cubic vessel.

The arrangement of the silo also has a major influence upon the explosion behavior. The maximum flame velocity increases with increased vent area A (Fig. 218), as anticipated. However, it is markedly lower for the vertical silo than for the horizontal arrangement due to the force of gravity.

The reduced maximum rate of pressure rise shows just the reverse behavior (Fig. 219). The values stemming from the horizontal silo are generally lower than

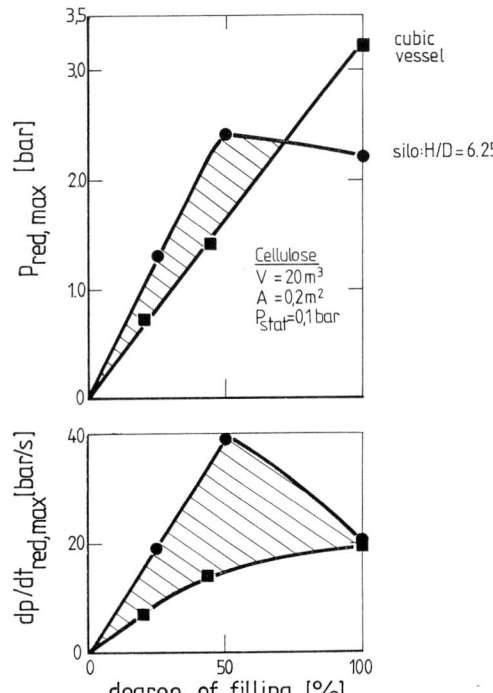

Fig. 217. Comparison of reduced maximum explosion characteristics at partial filling with dust/air mixture

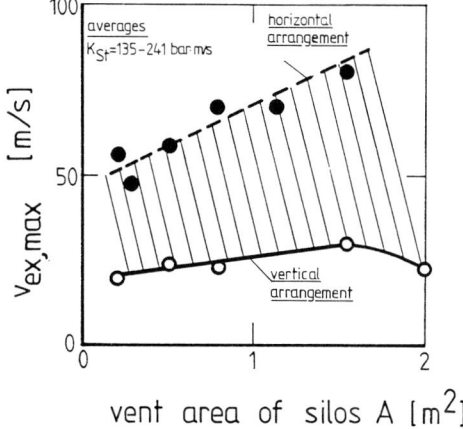

Fig. 218. Influence of the silo arrangement upon the flame velocity $v_{ex,max}$

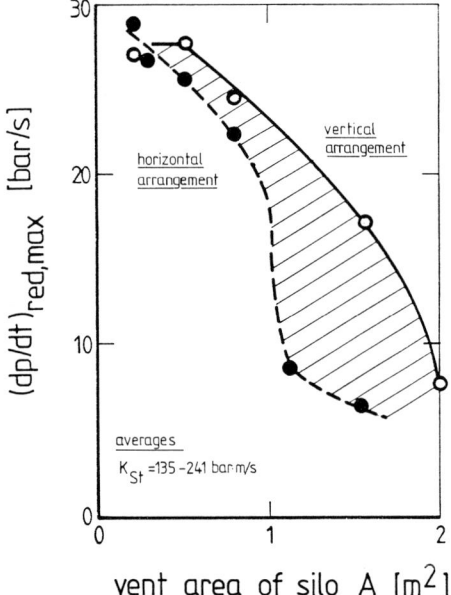

Fig. 219. Influence of the silo arrangement upon the reduced maximum rate of pressure rise $(dp/dt)_{red,max}$

the ones from a vertical arrangement. These are the reasons why different area requirement curves exist for every dust tested which depend upon the silo arrangement and the location of the ignition source. Figures 220 and 221 show the "composite area requirement curves" for wheat and corn starch dusts in a silo in comparison with the ones determined in a cubic vessel.

For small area venting ($A < 0.4$-m^2) the reduced maximum explosion pressure in the silo is lower than in the cubic vessel, independent of the dust type, because once the dust explosion is initiated, the flame reaches the cooling wall of the silo

5.3.3 Explosion Pressure-resistant Design with Explosion Pressure Venting

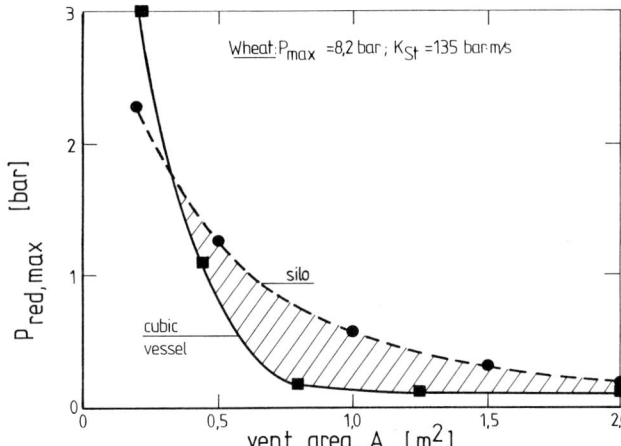

Fig. 220. Wheat: consolidated area requirement curves ($V = 20\text{-m}^3$, $p_{stat} = 0.1$ bar)

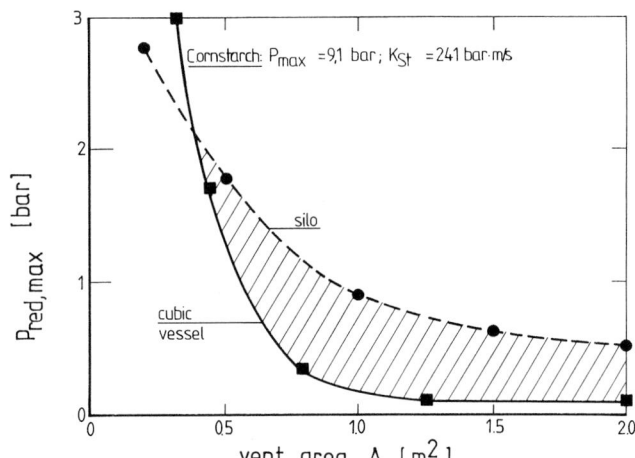

Fig. 221. Corn starch: consolidated area requirement curves ($V = 20\text{-m}^3$, $p_{stat} = 0.1$ bar)

much more rapidly. For large area venting ($A > 0.4\text{-m}^2$), however, the values of the reduced pressure are markedly higher in the silo than in the cubic vessel, the more so the higher the dust-specific characteristics.

Silos, therefore, have larger area requirements than cubic vessels. The demand of the VDI guideline 3673 "Pressure release of dust explosions" [55] to allocate the whole roof of a silo for the pressure release system is therefore justified. In such a case ($A = 2.0\text{-m}^2$), the pressure developed by the weaker-reacting wheat corresponds approximately to the static venting pressure of the venting system. For the more violently reacting corn starch, such pressure is 400% higher. This can only be accommodated by increasing the explosion resistance of the silo. In practice, it is necessary to design the vent openings of silos for pedestrian or vehicular traffic. This is often done by adding reinforced concrete covers loosely applied, with cast joints (Fig. 222) or with hinges (Fig. 223). The resulting relief capabilities (Fig. 196) have to be known for the selection of the explosion resistance of the silo.

Dust explosion tests with wheat have shown (Fig. 224) that reinforced concrete covers markedly increase the reduced maximum explosion pressure in comparison with a plastic membrane (Fig. 225). This is mainly due to the restricted venting process and not as much due to the increased static venting pressure of the relief arrangement. The venting efficiency is between 20% and 31% depending upon cover design.

The hinged, reinforced concrete cover (Fig. 223) performed perfectly, while the loosely applied and the solidly cast, reinforced concrete covers flew away in the case of an explosion, thereby presenting a danger for the surroundings. All statements presented so far have been based on a special dust dispersion procedure [55]. The real practical filling conditions have not yet been considered. A common filling method consists of pneumatic conveying. Therefore, the question was raised about what shape the area/pressure curves take once the dust/air mixture (e.g., corn starch) is pneumatically conveyed into the vertically arranged silo [116, 120].

Fig. 222. Solidly cast, reinforced concrete cover as the vent arrangement on top of a silo

5.3.3 Explosion Pressure-resistant Design with Explosion Pressure Venting

Fig. 223. Hinged, reinforced concrete cover as the vent arrangement on top of a silo

Fig. 224. Dust explosion test with a reinforced concrete cover as the vent arrangement on top of a silo

196 5 Protective Measures Against the Occurrence and Effects of Dust Explosions

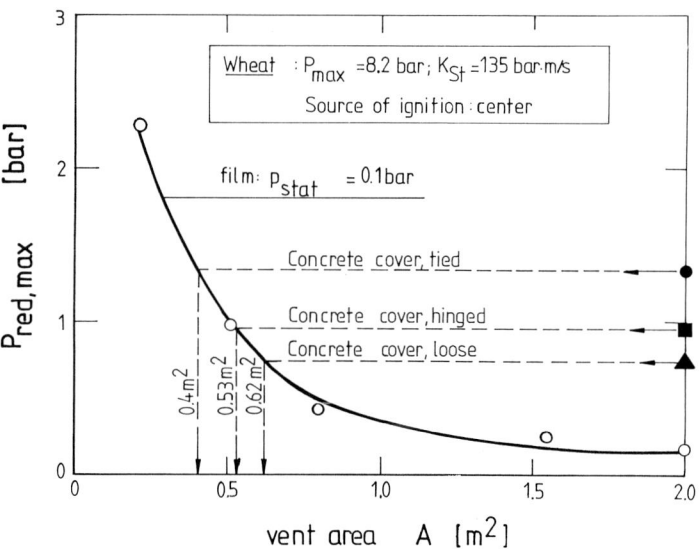

Fig. 225. Vent efficiency of reinforced concrete covers as the vent arrangement on top of a silo

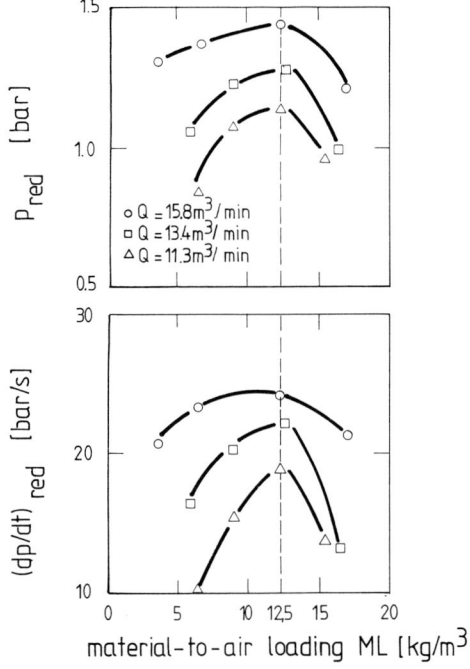

Fig. 226. Direct conveying of corn starch: Reduced explosion indices as a function of air volume Q and material-to-air loading ML (20-m³ silo; $A = 0.5$-m²; $p_{stat} = 0.1$ bar)

5.3.3 Explosion Pressure-resistant Design with Explosion Pressure Venting 197

At first, the direct conveying of the product into the silo will be analyzed. As shown in Fig. 226, there exists an optimum for the reduced explosion indices at an optimum material-to-air loading ML and a chosen air volume Q. The reduced explosion indices have a linear correlation with the air volume (Fig. 227). Through extrapolation, an air volume Q = 21 m^3/min corresponding to a feed velocity v = 45 m/s will result in the same values obtained using the VDI dust dispersion procedure. The optimum material-to-air ratio is independent of the chosen vent size. Figure 228 compares the pressure/area curves obtained in the silo using different dust dispersion procedures with the ones measured in a cubic vessel of equal size.

Except for very small vent sizes, the reduced maximum explosion indices obtained with pneumatic conveying are markedly higher than the ones obtained in cubic vessels, but slightly below the ones gained from using the VDI dust dispersion procedure in the silo. Therefore, the reduced explosion pressure ($p_{red,max}$ = 0.32 bar) reached with direct filling using the whole silo top as a vent area will be 167% higher than in the cubic vessel ($p_{red,max}$ = 0.12 bar). These figures are referenced to an air volume Q = 15.8 m^3/min. Once increased to approximately 21 m^3/min, the area requirement curve will coincide with the one valid for the VDI dust dispersion procedure. In that case, the pressure generated in the silo ($p_{red,max}$ = 0.51 bar) will be 325% higher than in the cubic vessel. This has to be considered when the explosion resistance is chosen for the silo to be proteced.

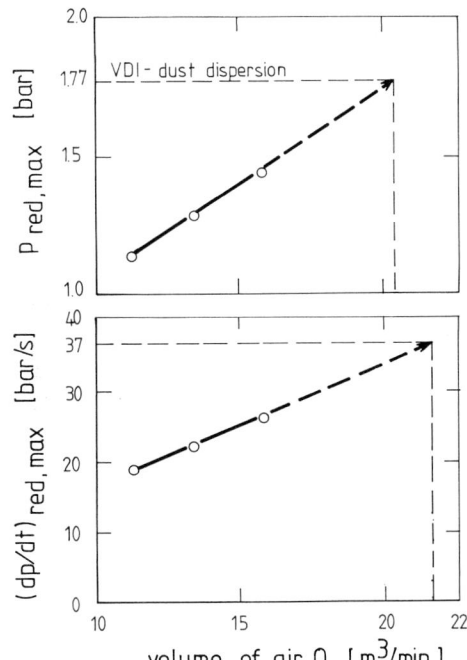

Fig. 227. Direct conveying of corn starch: Correlation of the reduced maximum explosion indices with the air volume Q and optimum material-to-air loading ML (20-m^3 silo; A = 0.5-m^2; p_{stat} = 0.1 bar)

198 5 Protective Measures Against the Occurrence and Effects of Dust Explosions

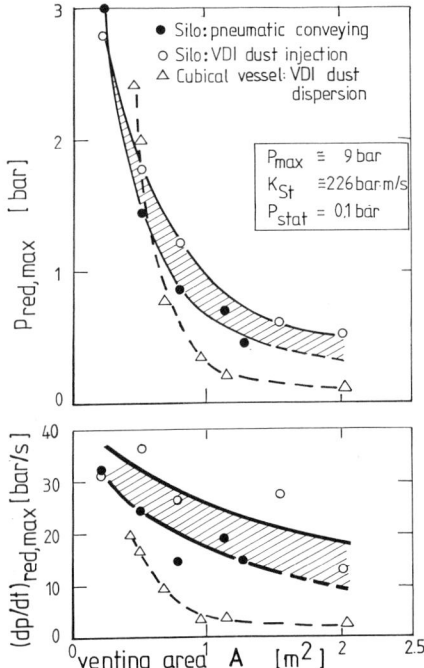

Fig. 228. Direct filling of corn starch: Influence of the dust dispersion procedure and the vessel shape upon the reduced maximum explosion indices relative to vent size A

Fig. 229. Corn starch explosion in the reinforced concrete silo. Filling from cyclone through rotary air lock (A = 1.27-m^2)

5.3.3 Explosion Pressure-resistant Design with Explosion Pressure Venting

If the product is fed from a cyclone through a rotary air lock into the silo (Fig. 229), then there will be a pronounced reduction in the reduced maximum explosion indices, in comparison with direct pneumatic filling.

Therefore, it is possible to show the influence the dust dispersion procedure has upon the occurrence of an explosion (Fig. 230).

The figure documents the already-stated differences between the VDI dust dispersion procedure and direct pneumatic filling. The first-mentioned procedure has some inherent reserves, provided that the air volume Q given in Fig. 226 and 227 is not exceeded. But it also shows the major reduction in explosion violence once the product is fed from a cyclone through a rotary air lock into the silo. The resulting measured reduced explosion indices correspond approximately to the ones obtained in a cubic vessel of equal size ($V = 20\text{-m}^3$) with the same relief area ($A = 1.27\text{-m}^2$) [$p_{red,max} = 0.22$ bar; $(dp/dt)_{red,max} = 3$ bar/s]. In using the outlined system instead of direct pneumatic filling, the required explosion resistance for the silo can be substantially lowered and costs saved. This is also confirmed by observations made on silos after dust explosions.

In this context, reference is to be made to yet unpublished test results for the determination of the lowest minimum ignition energy (LMIE) of hard-to-ignite granules with a few weight percent of fine particles in the 20-m³ silo. The minimum ignition energy required by the dust cloud generated by direct pneumatic filling is ten times the value of the LMIE of the fine dust because of extreme turbulence. If the product is filled into the silo from the cyclone through the rotary air lock, however, the minimum ignition energy of the dust cloud is identical with the LMIE of the fine dust because of the low turbulence.

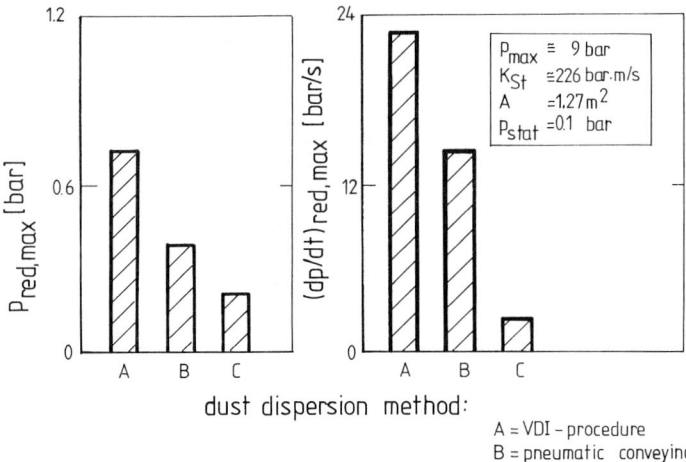

Fig. 230. Influence of the dust dispersion procedure upon the reduced maximum explosion indices (20-m³ silo, corn starch)

200   5 Protective Measures Against the Occurrence and Effects of Dust Explosions

In summarizing, the following can be stated: The dust cloud generated by direct pneumatic filling of the silo will be much harder to ignite with an extended capacitor discharge, and the reduced maximum explosion indices will be higher than the ones obtained when filling from a cyclone through an air lock.

5.3.3.4 Explosion Pressure Venting of Pipelines

The substantial damage from dust explosions in pressure-vented pipe systems raised some doubts with regard to the effectiveness of such systems. Thorough systematic testing was initiated in order to develop the most suitable design for pressure venting of pipelines [21] and to pin-point the most favorable location for installation. It became obvious that an effective pressure venting of the pipe system was only possible if safety membranes of ample size are installed within short intervals (1–2-m) on the pipe wall because of the directional influence of the explosion. In such cases, it is, however, imperative to consider the flames resulting from the explosion (Fig. 231), which may limit the use of such pressure-vented pipe systems to open air installations.

If a vent flap (door, panel) is used to close the vent opening of the pipeline (Fig. 232), the influence of the weight of the flap has to be considered, notably its effects upon the course of the explosion and also upon reduced maximum explosion indices (Fig. 233). The heavier the weight of the explosion door, the larger the explosion effects, e.g., in the case of a bucket conveyor [121]. The flame propagation in the vicinity of the vent opening is also substantial (Fig. 234).

If the arrangement of vent openings along the whole length of the pipeline is impossible, e.g., when installed in manufacturing buildings, then such pipe systems

Fig. 231. Wood dust explosion in pressure-vented pipeline (vent openings closed with safety membranes)

5.3.3 Explosion Pressure-resistant Design with Explosion Pressure Venting

Fig. 232. Element for a bucket elevator with explosion panel

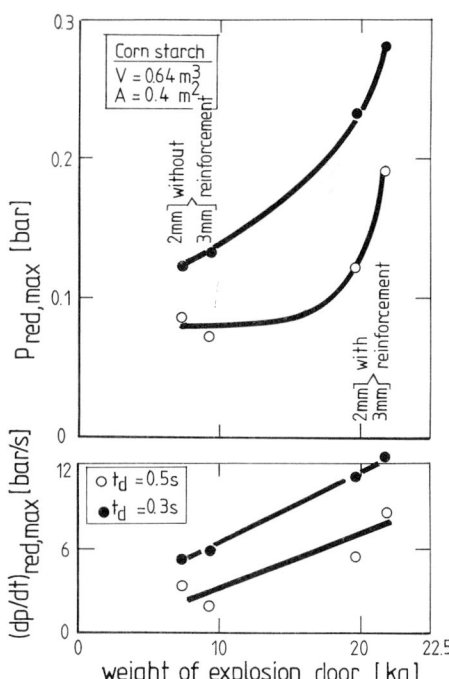

Fig. 233. Element for a bucket elevator. Influence of the panel weight upon the reduced maximum explosion indices
($t_d = 0.5$ s; $p_{max} = 7.8$ bar; $K_{St} = 157$ bar·m/s)
($t_d = 0.3$ s; $p_{max} = 8.4$ bar; $K_{St} = 217$ bar·m/s)

202   5 Protective Measures Against the Occurrence and Effects of Dust Explosions

Fig. 234. Element for a bucket elevator with explosion panel: Corn starch dust explosion

have to be built explosion-resistant for the expected dust explosion. Experience and test results indicate that the resistance of the pipeline, inclusive of flanged connections, is sufficient if designed for a nominal pressure of at least PN 10 for explosions starting at ambient pressure. Such pipelines will also withstand the short pressure peaks of detonation-like events or actual detonations. However, at end flanges and bends, substantially higher pressures will be encountered relative to the plain pipe due to pressure piling of the mixtures ahead of the flame front plus reflection (Fig. 137). Therefore, venting of such a location, without area restriction, is needed to protect such parts from destruction. This can be done either with rupture discs or other tested venting devices [explosion flap (Fig. 235), spring-loaded valve].

Moreover the activation of such a venting device by an explosion will accelerate the flame velocity due to pressure piling (see sect. 4.5), resulting also in a higher explosion pressure. This happens more readily the lower the activation pressure is. Therefore, the activation pressure has to be chosen at an accordingly high level ($p_{stat} = 0.5-2$ bar) in order not to favor detonation-like behavior or detonations as the result of venting. Venting devices which close on their own after venting shall only be used if their performance has been proven by explosion tests.

Fig. 235. Pressure venting device at dead end: Weighted explosion flap

5.3.4 Explosion-resistant Construction for Reduced Maximum Explosion Pressure in Conjunction with Explosion Suppression

Explosion suppression systems [122–126] are devices which prevent an unacceptably high pressure from dust explosions in vessels (which have not been designed explosion-resistant for the full maximum explosion pressure). They prevent destruction of the vessel and protect the operators working within the confines of the equipment area. Explosion suppression systems restrict the zone of influence of the explosion flame right at the start of an explosion. An often observed, subsequent work room explosion can be prevented in areas where dust settling cannot be prevented. Explosion suppression is important if an explosion of toxic or ecologically dangerous substances is anticipated.

However, one must assume that the explosion indices (maximum explosion pressure p_{max}; dust-specific K_{St}-value) are within range for the explosion suppression system. Explosion suppression is possible independent of the location of the equipment to be protected. However, the effort for maintenance will be greater (activation pressure of pressure sensors, check of driving media pressure and electronic amplifiers) in comparison with explosion pressure-vented systems. This may

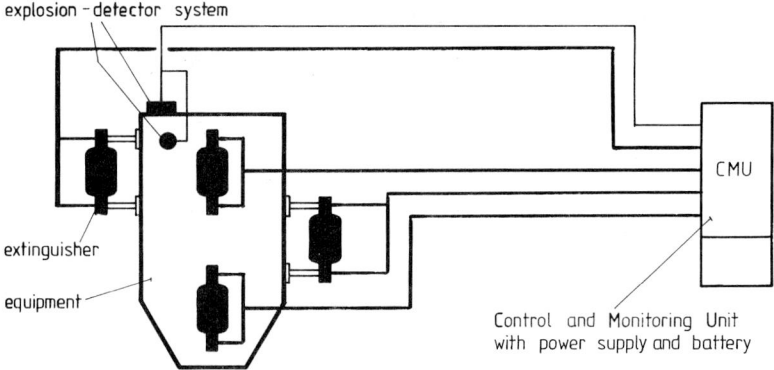

Fig. 236. Layout of an explosion suppression system

possibly be the reason why such a protective measure is only slowly being adopted by industry, regardless of its often-proven effectiveness in practice. Explosion suppression systems (Fig. 236) consist of a detector system which senses the start of an explosion, the pressurized container with the extinguishing media and a valve which is activated by the pressure sensor through a control-monitoring unit. The extinguishing media are dispersed into the protected vessel in as short a time as possible in order to reduce the expected maximum explosion pressure to a substantially lower level (Fig. 237). Explosion suppression systems have a built-in monitoring system and maintain their function for a given time after power failure.

The control and monitoring unit has the following assignment:
- start-up of the total explosion suppression system
- safety shut-down of the process equipment protected by the suppression system in case of an explosion
- power supply, either from main system or emergency battery
- monitoring charging current for emergency battery
- check of external wiring inclusive of ground fault monitoring
- automatic fault annunciating
- optical and acoustical alarm at activation of the explosion suppression system
- monitoring the pressure sensors for proper function.

The detectors which trigger the explosion suppression have to be capable of following the starting explosion sufficiently rapidly and without too much inertia. There are three types of detectors:

- thermoelectric detectors
- optical detectors
- pressure sensors

The first two types are unsuited for use with explosion suppression systems on vessels because they have to be installed close by the ignition source or else the dust

5.3.4 Explosion-resistant Construction with Explosion Suppression

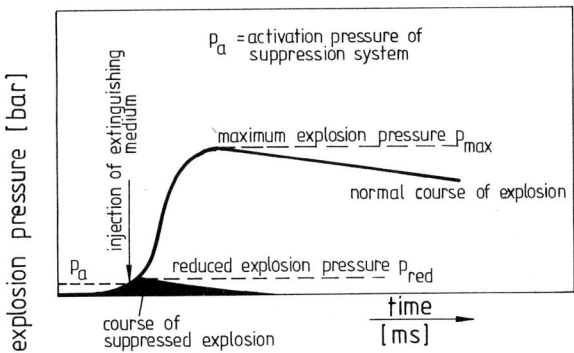

Fig. 237. Pressure behavior versus time for a normal and "suppressed" dust explosion

cloud which appears between the detector and the ignition source will absorb some of the radiation. Such action will delay the signal and result in too late an activation of the suppression system.

Since the explosion pressure of a starting explosion propagates uniformly at the velocity of sound in the vessel, the pressure sensor is the most suitable device. Such sensors (Fig. 238) register with ample safety a starting explosion at a pre-set pressure and can trigger via the control unit the ignition impulse for the valve of the extinguishing container. They have proven themselves amply in industrial practice.

The pressure sensors which activate the explosion suppression must be insensitive to the vessel contents and especially to external influences (shock, vibration). Therefore, they are normally used in pairs, i.e., two sensors offset by 90° have to

Fig. 238. Membrane pressure sensor for explosion suppression systems

reach simultaneously the activation pressure for the system. This should also be verifiable. The pressure sensors have to be deactivated before a vessel entry takes place and they must be secured against accidental tripping. The extinguishing containers must fulfill all rules and regulations applicable for the user and be designed in accordance with the technical guidelines.

The containers are equipped with fast-acting valves which will open the full cross section within milliseconds after the activation signal is given. They are designed in such a way that the whole content of extinguishing medium will be uniformly distributed in the protected vessel within as short a time as possible. Such requirements are met by the extinguishing containers shown in Fig. 239 and 240, which can be described as:

- 5-l container with rise pipe and two 3/4" valves activated by blasting detonators
- 5-l container with a 3" valve activated by a line-cutting charge.

Both containers are filled with 4 kg of extinguishing medium. The large amount of driving medium (nitrogen) provides an optimum dispersion of the extinguishing medium inside the protected vessel.

The pressure of the driving medium is as follows:

3/4" system $p_{N_2} = 120$ bar

3"　system $p_{N_2} = 60$ bar

a

b

Fig. 239 a/b. 5-l extinguisher with two 3/4" valves.
a) container; b) hemispherical nozzle for dispersion of the extinguishing medium

5.3.4 Explosion-resistant Construction with Explosion Suppression

Fig. 240 a/b. 5-l extinguisher with a 3″ valve **a)** container **b)** hemispherical nozzle for dispersion of the extinguishing medium

Through the special arrangement of the nozzles (Fig. 239 and 240), a uniform distribution of the extinguishing medium is guaranteed. Special attention has to be given to the proper functioning of the hemispherical nozzles, which must not be affected by operating conditions, especially product accumulations.

The hemispherical nozzles which distribute the extinguishing medium penetrate into the protected vessel in the suppression arrangements described so far. This is often unacceptable for practical applications, especially if the equipment which processes combustible dusts is subjected to frequent product changes. For such cases, movable hemispherical nozzles or so-called telescopic nozzles (Fig. 241) are to be recommended. At first, the hemispherical nozzle is outside the vessel to be protected and separated by a membrane. In case of an explosion, i.e., once the fast-acting valve of the extinguisher is activated, the nozzle is propelled forward by the pressure of the driving medium, destroying the membrane and initiating the extinguishing action. The mechanical process results in a delayed start of the extinguishing, which in turn causes an increased reduced maximum explosion pressure in the vessel (which generally is within an acceptable range). The connecting lines between the valve of the extinguisher and the hemispherical nozzle which disperses the extinguishing medium has to be as short as possible and sufficiently pressure shock-resistant.

In case longer lines are needed because of technical reasons (especially in conjunction with the 3/4" systems), then the length should not exceed 500-mm, otherwise the extinguishing efficiency will be reduced.

Both explosion suppression systems have practically the same extinguishing efficiency using a powdery extinguishing medium for incipient dust explosions. For

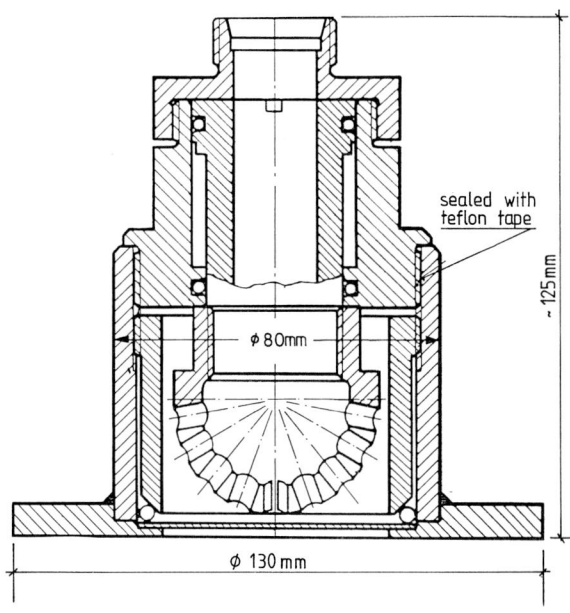

Fig. 241. Telescopic hemispherical nozzle for the ¾" suppression system

5.3.4 Explosion-resistant Construction with Explosion Suppression

a given combustible dust, there exists basically a correlation between the activation pressure of the suppression system and the reduced maximum explosion pressure $p_{red,max}$, as shown in Fig. 242. The reduced maximum explosion pressure is the pressure at optimum combustible dust concentration which will exist in the protected vessel after the activation of the extinguishing system. A rising activation pressure p_a results also in an increased reduced maximum explosion pressure $p_{red,max}$. An extinguishing medium will be classified as "very good" if an increase in the activation pressure will result in a minimal rise in the reduced maximum explosion pressure. Only extinguishing agents with the best effectiveness fulfill such a requirement.

In general, a dust explosion may be considered successfully suppressed if the maximum explosion pressure p_{max} is lowered to a reduced maximum explosion pressure $p_{red,max} = 1$ bar, with an activation pressure $p_a = 0.1$ bar. This means that the vessel and the equipment which are protected by explosion suppression have to be built explosion-resistant for a 1 bar range in accordance with paragraph 5.3.1. The following extinguishing agents are used in practice:

– halogenated hydrocarbons
– water
– extinguishing powders, and most recently
– hybrid extinguishers.

Basically, it is also possible to use halons in conjunction with explosion suppression systems for fighting dust explosions. They are liquid extinguishers which will evaporate in the protected vessel once discharged from the container. The prerequisite for halon use is that the extinguishing agent be dispersed very quickly after the start of ignition. This calls for a relatively low activation pressure ($p_a < 0.1$ bar) in order to have a successful explosion suppression. This unfortunately may lead to accidental triggering because of pressure fluctuations in the ongoing process. Also, some shortcomings became known which may make the application of halon suppression systems problematical. If, for instance, halon is dispersed too late into the ves-

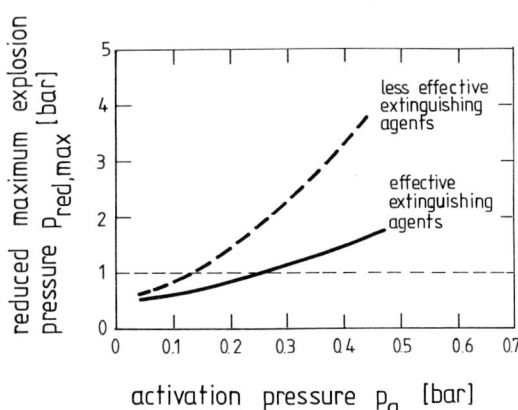

Fig. 242. Effectiveness of explosion suppression systems

210   5 Protective Measures Against the Occurrence and Effects of Dust Explosions

sel because of an elevated activation pressure or if the system is underdesigned, then the explosion may react much more violently because of the decomposing halon.

Water has proved itself as an effective suppressant for explosions of grain- and feed-dusts.

Extinguishing powders have a good suppression effectiveness if they are suitable for inerting once admixed with combustible dusts (see sect. 5.2.3.1.5). In this context, extinguishing agents with ammonium phosphate as a base have to be mentioned. With a suitable pressure resistance of the vessel, there is quite some margin in selecting the activation pressure. Even with relatively high activation pressures ($p_a > 0.1$ bar), there is still a substantial damping of the course of the explosion in the protected vessel. The disadvantages mentioned for halon cannot be observed.

Hybrid extinguishers consist of a particular halon mixed with a specific powder. Their advantage lies in the inerting effect halon takes after a successful explosion suppression in the protected vessel. At this time, no experimental experience is available in Europe with such an extinguishing agent. Basically, it is necessary to determine the suitability of the extinguishing agent for the given application with explosion suppression tests [126] in large enough vessels ($V \geq 1\text{-m}^3$). At the same time, the correlation between activation pressure of the suppression system and the reduced maximum explosion pressure has to be established. In addition, the suppressant has to be insensitive to the temperature and vibration inherent in the system to be protected.

In designing the explosion suppression system for the protection of vessels, the following question has to be answered:

How much extinguishing material is required in order to effectively suppress an explosion of the given combustible dust in a known vessel size? An answer is possible based on the results obtained from systematic tests (vessel size 1-m^3 to 60-m^3) with the assumption that:

– the extinguishing agent be in a 5-l container (Fig. 239, 240) at the proper driving medium pressure
– the most suitable extinguishing agent be used.

The test results are shown in Fig. 243. It was possible to suppress explosions of quiescent propane/air mixtures as well as turbulent dust/air mixtures of the dust explosion class St 1 in vessels up to 60-m^3 with a given amount of extinguishing medium (number N of 5-l containers). These tests used an activation pressure for the suppression system of $p_a = 0.1$ bar, and the reduced maximum explosion pressure was lowered to below 0.5 bar. From Fig. 243 it can also be seen that the extinguisher requirements do not follow proportionally the size of the vessel to be protected but are determined by the cubic law. Therefore, the following equation can be applied to calculate the required extinguishing media (N_2) for a vessel size (V_2) with known values (N_1) for a vessel size (V_1):

$$N_2 = N_1 \cdot \frac{\sqrt[3]{V_1}}{\sqrt[3]{V_2}} \cdot \frac{V_2}{V_1} \quad \left[N = \frac{\text{number of 5--l}}{\text{extinguishers}} \right]$$

5.3.4 Explosion-resistant Construction with Explosion Suppression

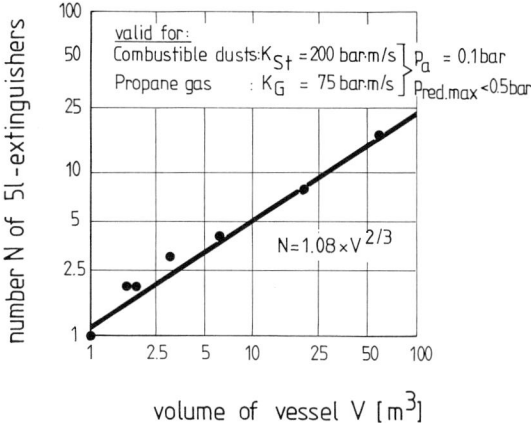

Fig. 243. Suppressant requirements for the explosion suppression of propane and dusts

If the requirements are known for a 1-m³ vessel then:

$$N_2 = N_1 (V_2)^{2/3}.$$

Table 22 gives the minimum number of 5-l extinguishers required for the protection of a vessel size V_2.

The number of extinguishers which is calculated from Table 22 has to be rounded to the next higher figure. The extinguishers also have to be distributed as evenly as possible on the surface of the vessel to be protected. The figures in Table 22 indicate some limitation for the use of an explosion suppression system for dust explosions. This limit is set by the dispersion velocity of the extinguishing agent of a multiple of 10 m/s. Based on present knowledge, suppressant systems can be effectively used for the protection of vessels against dust explosions, if the K_{St}-value does not exceed 300 bar·m/s. Dusts having a greater explosion violence (aluminum dust $K_{St} > 300$ bar·m/s) may not be effectively suppressed in vessels over the whole range of concentration. At the most, the protective system may restrict the explosion limits up to a dust concentration which will result in explosion indices which remain below the stated limit for application.

Table 22. Minimum requirements of suppressants for an explosion suppressant system ($p_a \leq 0.1$ bar, best-suited extinguishing medium)

Dust explosion class	Number of 5-l extinguishers N_2	Minimum explosion resistance of the vessel [bar]
St 1	$1.08 \cdot V_2^{2/3}$	0.5
St 2	$1.40 \cdot V_2^{2/3}$	1.0

212 5 Protective Measures Against the Occurrence and Effects of Dust Explosions

In applying the above-outlined system with 5-l extinguishers for the protection of 100-m^3 or 200-m^3 vessels, between 23 and 48 containers are required, depending upon vessel size and dust explosion class. Not only the number of extinguishers but also their installation will cost considerably. Therefore, systematic tests were made in vessels of 10-m^3, 25-m^3, and 250-m^3 in order to improve the effectiveness of the original suppression system [127]. The size of the container as well as the diameter of the valve with the line-cutting charge were increased systematically (Fig. 240). Such a developement culminated in a 45-l extinguisher with a 5-inch valve (Fig. 244). The capacity is 35 kg of extinguishing agent on the basis of ammonium phosphate, which uses nitrogen as the driving medium at a pressure of $p_{N2} = 60$ bar. The explosion suppression tests were successfully concluded in a 250-m^3 vessel (Fig. 245). The results are shown in Fig. 246.

Also, the large vessel shows the expected correlation of activation pressure p_a of the improved suppression system with the reduced maximum explosion indices ($p_{red,max}$; $(dp/dt)_{red,max}$, respectively). A successful suppression of a dust explosion requires 7–10 45-l extinguishers, depending upon the dust explosion class. The reduced maximum explosion pressure remains below $p_{red,max} = 1$ bar even with a relatively high activation pressure $p_a = 0.2$ bar. In using the original suppression system, at least 43–56 5-l extinguishers would have been required for the same suppression result. The new, improved explosion suppression system allows a substan-

Fig. 244. 45-l extinguisher with 5″ valve

5.3.4 Explosion-resistant Construction with Explosion Suppression

Fig. 245. 250-m³ vessel equipped with dust cyclinders and 45-l extinguishers

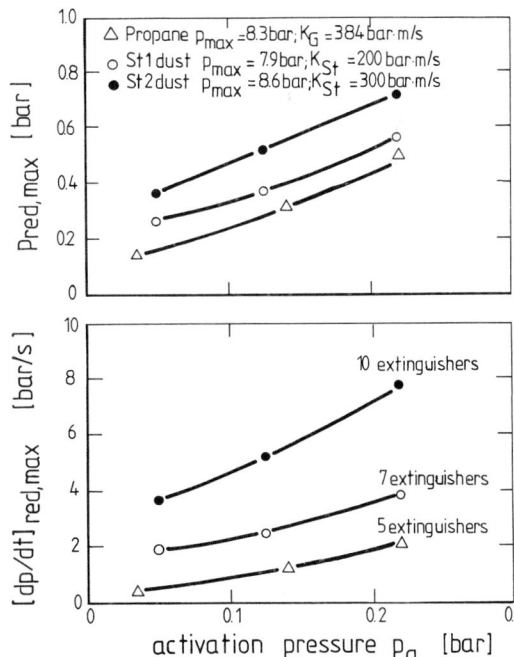

Fig. 246. 250-m³ vessel. Explosion suppression. Correlation of activation pressure p_a with reduced maximum explosion indices

214   5 Protective Measures Against the Occurrence and Effects of Dust Explosions

tial decrease in extinguishers for the fighting of dust explosions. As shown also in Fig. 246, the suppression of a propane explosion requires less extinguishers than a dust explosion. This is due to the course of the explosion. In the large vessel, the explosion pressure of the ignited, turbulent dust/air mixture increases at first slowly and then reaches the final pressure quite rapidly.

For the propane/air mixture, however, which is ignited in a quiescent state, at least 600 ms pass until a noticeable pressure is detected which will trigger the suppression. If this does not happen, then another 400 ms pass until a rapid pressure rise occurs towards the end of the combustion, which will result in the high gas-specific K_G-value.

The results from these systematic explosion suppression tests in vessels having varying sizes again show that the well-known cubic law determines the extinguisher requirements per vessel size. Figure 247 documents that the improved explosion suppression system allows a 80% reduction in the number of extinguishers needed, based on the customary system to successfully fight dust explosions. Thus, the application of the explosion suppression system is not only made easier, especially for larger vessels ($V \geq 10\text{-m}^3$), but also the installation cost is reduced.

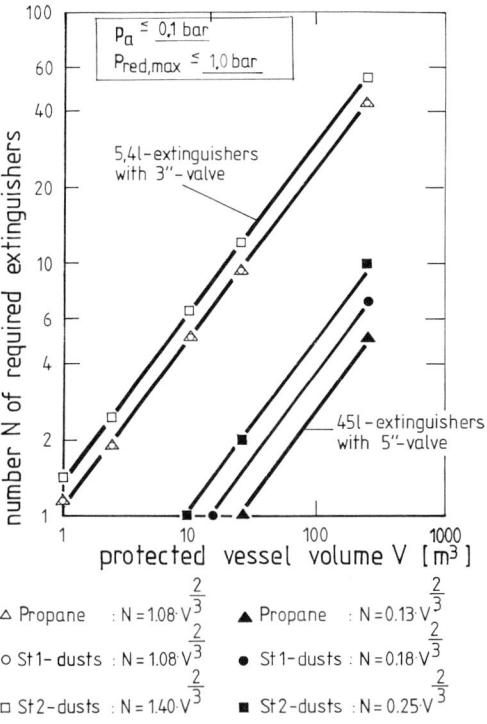

△ Propane   : $N = 1.08 \cdot V^{\frac{2}{3}}$ ▲ Propane   : $N = 0.13 \cdot V^{\frac{2}{3}}$

○ St1- dusts : $N = 1.08 \cdot V^{\frac{2}{3}}$ ● St1- dusts : $N = 0.18 \cdot V^{\frac{2}{3}}$

□ St2- dusts : $N = 1.40 \cdot V^{\frac{2}{3}}$ ■ St2- dusts : $N = 0.25 \cdot V^{\frac{2}{3}}$

Fig. 247. Suppressant requirements for explosion suppression systems for vessels. Usual system versus improved system

5.3.5 Technical Diversion or Arresting of Explosions

5.3.5.1 Preliminary Remarks

Diversion devices are always necessary in systems which are made up of equipment using preventive explosion protection (see sect. 5.2) and protection by design measures (see sects. 5.3.1–5.3.4). In the latter, effective ignition sources are anticipated which may result in a dust explosion [128] that has to be isolated from the other equipment. Such devices may also be required if a dust explosion is expected to propagate out of a long pipeline into equipment which is protected by design measures. Despite such protection, the flame jet ignition of the dust/air mixture (perhaps with pressure piling) may damage or even destroy said equipment (Fig. 248).

Fig. 248. Explosion pressure-vented filter housing (V = 13.6-m^3, A = 3-m^2) with flame jet ignition

5.3.5.2 Extinguishing Barrier

Mechanical flame arresters with ribbon inserts (Fig. 249), which are used against the propagation of flammable gas or vapor explosions, especially in pipelines, are not suited for dust explosions because of the tendency to plug the small areas.

Therefore, the automatically operating extinguishing barrier (Fig. 250) was developed for dust-carrying pipelines [129, 130]. It can stop fully-developed dust explosions at a predetermined pipe location and limit the course of the explosion to a defined pipe section. The advantage of such a type of arrester in comparison with the usual mechanical flame arresters is the prevention of the always inconvenient

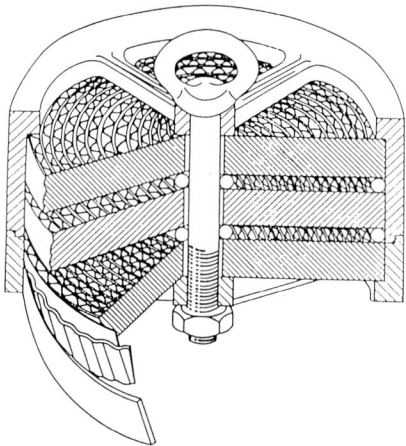

Fig. 249. Schematic presentation of a triple flame barrier with ribbon inserts

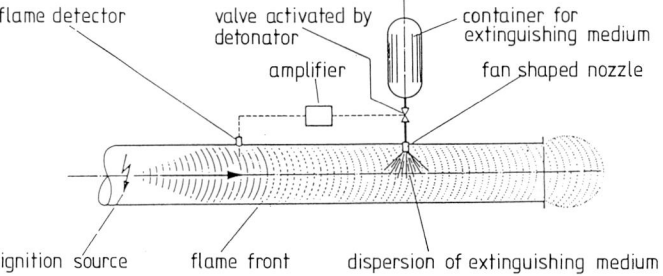

Fig. 250. Schematic presentation of an automatic extinguishing barrier

pressure drop in pipelines. The effectiveness of the barrier hinges upon the detection of the dust explosion in the pipeline with a suitable detector. Its signal is amplified by the control unit (Fig. 236), which will open the fast-acting valves of the extinguishers, thereby freeing the path for the extinguishing agent, which will be dispersed inside the pipeline through a nozzle arrangement. An optical flame detector (Fig. 251) is the most suitable device for such a barrier arrangement since the propagating flame from the explosion has to be detected and extinguished. Due to the intense flame which is generally found in dust explosions, one can use daylight-insensitive flame detectors. A possible penetration of daylight into the pipe will therefore not cause a false activation.

It is necessary to flush the optical lens with gas in order to keep it dust-free. Pressure detectors are not suited for the case on hand because there is no distinct separation between the pressure and flame fronts for explosions in pipelines.

5.3.5 Technical Diversion or Arresting of Explosions

Fig. 251. Infrared flame detector

To store the extinguishing agent, the same containers are used as for explosion suppression (Fig. 239):

- 5-l extinguisher with one 3/4″ valve, filled with 2 kg of the most suitable agent; used in pipelines with smaller cross sections (DN ≤400 mm)
- 5-l extinguisher with two 3/4″ valves, filled with 4 kg of the most suitable agent; used for pipes with larger cross sections (DN ≥500 mm)

As extinguishing agent, a powder based on ammonium phosphate is preferred. The containers are pressurized with nitrogen as the driving medium at $p_{N2} = 120$ bar. The use of halons gives the same adverse effects as described in conjunction with the safety measure explosion suppression. Vertically arranged fan-shaped nozzles are used for the dispersion of the extinguishing agent (Fig. 252).

Intensive studies have shown (Fig. 253) that the suppressant requirement at the barrier location must be adjusted linearly with the flame velocity. Especially at high velocities, the extinguishers with the 4 kg content are more effective than the ones with 2 kg. For a given flame velocity, the required amount of extinguishing agent per cross section (m^2) is constant in order to force the disruption of the explosion, independent of the pipe diameter. Based upon this, the figures for Table 23 have been generated giving the required number of extinguishers per flame velocity v_{ex} and nominal pipe diameter DN.

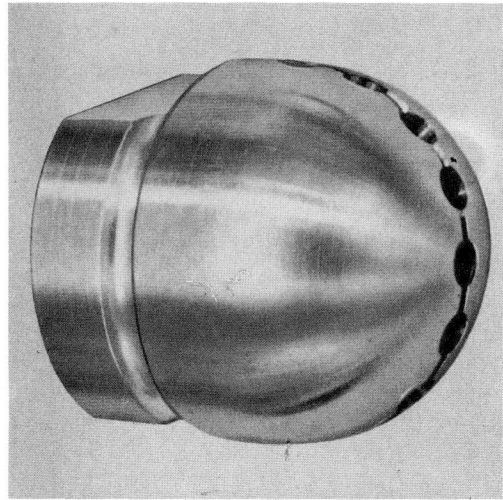

Fig. 252. Fan-shaped nozzle for the dispersion of extinguishing powder for extinguishing barriers

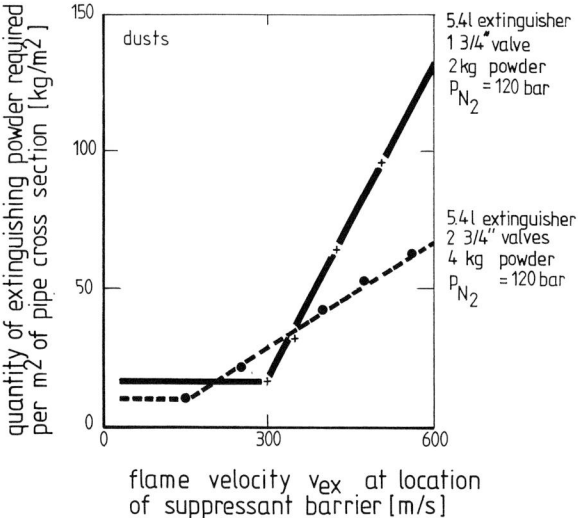

Fig. 253. Influence of the flame velocity at the barrier location upon the required suppressant per cross section (m²) (Pipeline DN 400, length 20-m, flame detector at 1-meter mark, flame barrier at 10-meter mark)

5.3.5 Technical Diversion or Arresting of Explosions

Table 23. Extinguishing barrier: minimum requirement of extinguishers (length of pipe: 20-m, flame detector at 1-meter mark)

DN [mm]	V_{ex} at barrier [m/s] 5-l extinguisher with	300	400	500	600
			Required number of extinguishers		
200		1	1	2	2
300	one 3/4" valve	1	2	4	5
400		1	4	6	8
500		2	2	3	4
600	two 3/4" valves	2	3	4	5
700		3	4	6	7

For the design of such flame barriers, the following may serve as a guideline: In a 200-m-long pipeline having a 400-mm diameter, the flame velocity at the 10-m mark, which is the location of the barrier, can be described as follows:

Dust explosion class

St 1 ($K_{St} \leq 200$ bar·m/s) : $V_{ex} \leq 300$ m/s
St 2 ($K_{St} = 201\text{–}300$ bar·m/s) : $V_{ex} = 301\text{–}400$ m/s
St 3 ($K_{St} > 300$ bar·m/s) : $V_{ex} > 400$ m/s

For small pipe areas, the values for the flame velocity move upwards; for large areas, downwards. Figure 254 shows the explosion indices for combustible dust in a 20-m-long pipeline with 400-mm diameter for an unhindered and an extinguished explosion. The estinguishing barrier is located at the 10-m mark, with the flame detector at the 1-m mark.

As can be seen, the propagation of the explosion flame is not only limited to the zone between ignition source and barrier, as expected, but very violent explosions or detonation-like behavior are stopped from occurring downstream of the barrier.

Figure 255 shows an actual installation of an extinguishing barrier in a vent pipe.

The use of extinguishing barriers is by no means limited to pipes with small cross sections. Applications with pipelines having diameters of 1400-mm–2500-mm (Fig. 256) have shown that there are no problems in fighting coal-mine explosions [methane explosions (Fig. 257), coal dust explosions, methane/coal dust explosions] and also explosions with dusts belonging to the dust explosion classes St 1 and St 2.

The above-mentioned tests have shown, on one hand, that the required quantity for the extinguishing agent in large area pipe changes proportionately with the expected flame velocity. Because of the proportional relationship between the flame velocity and the explosion pressure (see sect. 4.5), there is, on the other

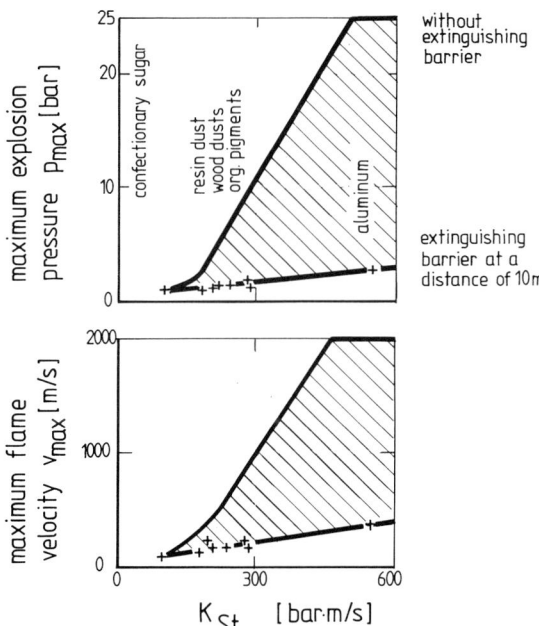

Fig. 254. Pipeline DN 400, lp = 20-m: Effectiveness of an extinguishing barrier against dust explosions

Fig. 255. Extinguishing barrier in a vent pipe

5.3.5 Technical Diversion or Arresting of Explosions

Fig. 256. Extinguishing barrier for pipelines with large diameter

hand, a linear correlation between expected explosion pressure and extinguishing requirement per cross section (m²). The requirement will be such that the explosion be successfully terminated at the location of the barrier (Fig. 258).

Extinguishing barriers are suited to stop dust explosions in bucket elevators and to prevent the explosion from entering an e. g., connected silo. As per statistics (Fig. 6), approximately 10% of all dust explosions occur in such installations.

The effectiveness of extinguishing barriers was tested in a reinforced 30-m-high belt and bucket elevator with pipe-casings (diameter D = 390-mm; Fig. 259). In the direction of flow immediately after the boot and before the head there was a barrier [consisting of two 5-l extinguishers facing each other with two valves, each activated by detonators (Fig. 260)]. The extinguishers were activated by flame detectors mounted in the wall of the casing [131, 132].

The tests were always made with the belt running and with feedstuff dusts (p_{max} ≤9.2 bar, K_{St} ≤200 bar·m/s, LMIE ≥10 mJ) under the following conditions:

- product in the buckets
- product in the buckets and additional explosible dust/air mixtures
- no product in the buckets but explosible dust/air mixture.

The additional dust/air mixture which was used for the tests was again prepared and introduced as usual through rapid dispersion of dust from containers into the system (feed and return casing), varying the dust concentration over a wide range.

Fig. 257. Methane explosion in pipeline DN 2500 (400-m³ mixture)

The optimum explosion violence was reached once ignition was initiated in the middle of the feed casing at the 15-m mark. The explosion developed fastest in the feed direction, i.e., the extinguishing barrier located in the head of the elevator was actuated first.

Due to their high content of large particles (M = 1000μm) and small content of fines 3 wt% <63 μm), unground soybean-extraction grouts are not explosible. But after conveying the product for several minutes, the fines were liberated and a self-propagating explosion noticed. The same observations were made by pneumati-

5.3.5 Technical Diversion or Arresting of Explosions

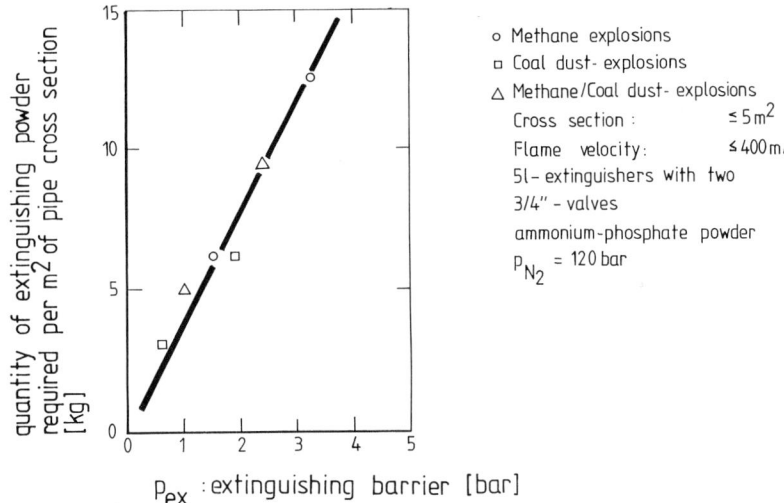

Fig. 258. Extinguishing barrier: correlation of extinguishing powder requirement with explosion pressure anticipated at the barrier location

Fig. 259. General view of the reinforced bucket elevator

224   5 Protective Measures Against the Occurrence and Effects of Dust Explosions

Fig. 260. Reinforced boot with extinguishing barrier (top)

cally conveying similar products into silos. The explosion violence which resulted from adding a dust atmosphere to the product in the buckets was always in between the results gained from the other two conditions. Therefore, these test results will be excluded from the following discussions.

Optimum explosion characteristics are always reached in the feed casing (Fig. 261), with an explosible atmosphere present, but without product in the buckets. The existence of the two extinguishing barriers substantially reduces both explosion characteristics, especially the maximum flame velocity in comparison with the unrestricted course of the explosion. The pressure build-up in the boot is markedly higher than in the head (Fig. 262) because of the longer distance the explosion takes through the return casing. Such pressure again is markedly reduced by the extinguishing barriers.

In general, the extinguishing barriers performed very well in all explosion tests. They not only reduced the explosion characteristics in the various parts of the elevator, but also prevented an explosion propagation into the return casing. The test results lead to the conclusion that bucket elevators which are protected with extinguishing barriers have to be built explosion pressure shock-resistant for 3 bar. This obviously corresponds to an explosion pressure resistance of 2 bar. If the elevator height exceeds 30-m, then additional barriers have to be installed every 30-m, with optical flame detectors upstream and downstream of the barrier. In practice, obvi-

5.3.5 Technical Diversion or Arresting of Explosions

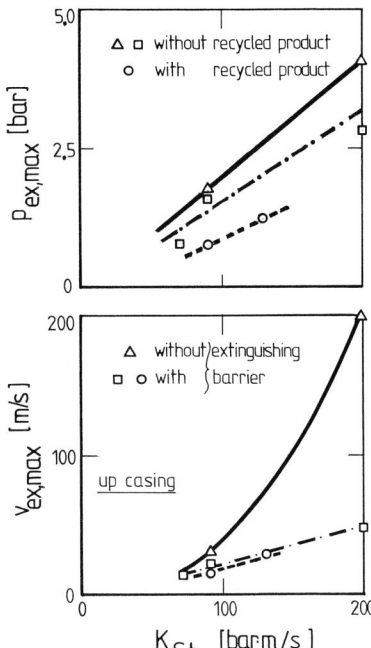

Fig. 261. Influence of the extinguishing barrier upon the explosion characteristics in the feed casing

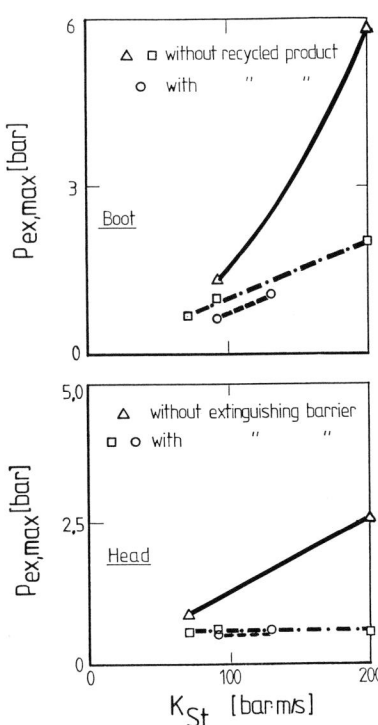

Fig. 262. Influence of the extinguishing barrier upon the maximum explosion pressure in the elevator boot and head

ously the return casing has to be protected in the same manner. The number of 5-l containers which is required for the design of extinguishing barriers can be taken from Fig. 263 in relation to the casing diameter. Figure 264 schematically depicts the layout of such barriers on a protected belt-and-bucket elevator.

In practice, a structure may be on top of the silo for weather protection. The floor of said structure must be suitable for pedestrian and vehicular traffic for silo inspection and cleaning. Therefore, it is generally not possible to route the vent ducts from the explosion vent in the top of the silo to a safe space (Fig. 7). Tests with an imitated vented structure on top of a pressure-vented silo (Fig. 265) proved that such endangering can be eliminated by the use of HRD extinguishers (**high rate discharge**) in the proximity of the silo vent (Fig. 266).

Three 10-l extinguishers have been used, each equipped with two detonator-activated valves. The extinguishers are arranged at 120° positions and are filled with 7 kg of extinguishing agent on the basis of ammonium phosphate (propellant pressure p_{N2} = 120 bar). The extinguishing agent is again dispersed through fan-type nozzles. The system is activated by an optical flame detector mounted in the silo wall. With such an arrangement, which is also some kind of an extinguishing barrier, it was possible to successfully extinguish the flames in the area of the vent and prevent the explosion from propagating into the structure. The flame originated from a corn starch explosion inside the silo at optimum dust concentration.

The explosion pressure inside the silo was $p_{red,max}$ = 0.5 bar, with a flame velocity v_{max} = 250 m/s. The pressure inside the top structure was only $p_{red,max}$ = 0.06–0.08 bar. Without extinguishers in service, the structure was destroyed at a pressure of $p_{red,max}$ = 0.3 bar. From this, it was shown that due to the exhaust of unburnt

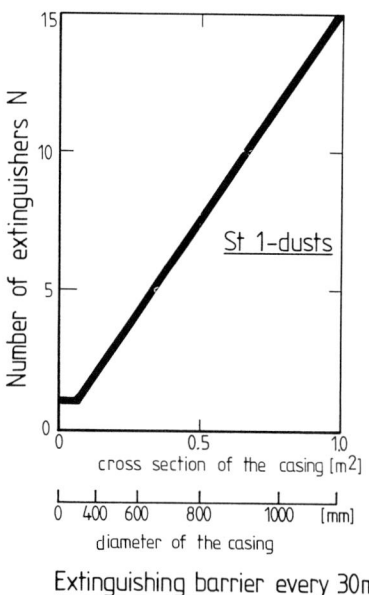

Fig. 263. Requirements for extinguishing barriers as a function of casing diameter (St 1 dusts, explosion pressure shock resistance: 3 bar)

5.3.5 Technical Diversion or Arresting of Explosions

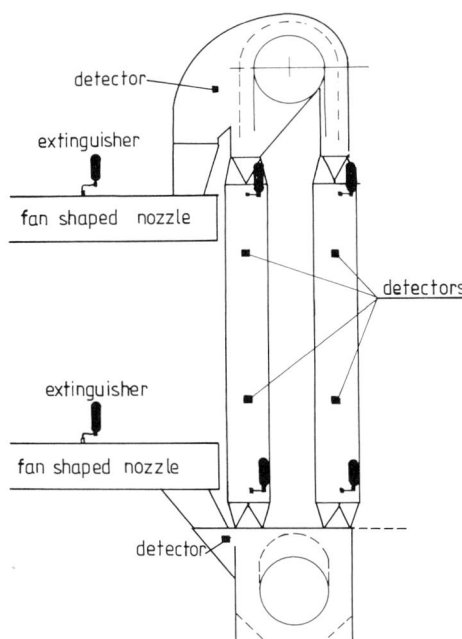

Fig. 264. Layout for extinguishing barriers on a protected belt-and-bucket elevator [132]

Fig. 265. Reinforced concrete silo (V = 20-m^3, A = 1.54-m^2) with vented structure (V = 100-m^3, A = 9-m^2)

Fig. 266. HRD extinguishers inside the structure close to the vented silo top

dust/air mixture into the structure ahead of the flame front, a larger silo height will be simulated. Figures 267 and 268 convey an impression of these tests.

The above statements have shown that extinguishing barriers have an outstanding effectiveness when used in pipelines (especially with large cross sections), in

Fig. 267. Exhaust of extinguishing agent from the vented structure with the HRD system

Fig. 268. Flames leaving the vented structure without HRD system

elevators, and when used to prevent a silo explosion from propagating into a structure. Such a barrier not only limits the course and effects of an explosion, but also markedly restricts the flame propagation.

5.3.5.3 Rotary Air Locks (Rotary Valves)

If combustible dust is discharged through a rotary air lock (Fig. 269) from equipment or a vessel protected with design measures, then such a lock may act as a mechanical flame barrier against dust explosions, provided certain design features are incorporated.

Early information about the propagation of ignition through narrow gaps comes from the recent, not yet concluded tests made by Schuber [133]. He is using a 42-l ignition chamber for the start of a dust explosion. A circular gap with a given length lg and width w connects the ignition chamber with the surrounding free area of a second 1-m^3 housing. The dust/air mixtures required for the tests are generated in both housings by rapidly discharging the dust from pressurized dust storage chambers (Fig. 270).

Fig. 269. Rotary air lock

Fig. 270. Test arrangement for the determination of the ignition propagation of dust/air mixtures through narrow gaps

Schuber found that for dust/air mixtures:

1) a low turbulence in the ignition chamber as well as in the second housing generally led to an easy propagation of ignition and
2) for a constant gap length lg, the maximum experimental safe gap w (MESG), which just did not propagate an explosion, was in the same range as that for the flammable gases methane or propane but could also be substantially larger.

5.3.5 Technical Diversion or Arresting of Explosions

In a double logarithmic plot, the gap width w is a linear function of the product lowest minimum ignition energy (LMIE) times reduced ignition temperature TI (Fig. 271). This means that both characteristics which influence the ignition behavior of extended capacitor discharges and hot surfaces also affect the ignition propagation of dust/air mixtures. Dusts having a low ignition temperature and LMIE (e.g., Lycopodium) also require a small maximum experimental safe gap w. If said characteristics have a large value (e.g., coal) then the maximum experimental safe gap will be accordingly large.

Figure 272 shows the influence of the gap length lg upon the maximum experimental safe gap w. Obviously, such a value does not change proportionately with the gap length, as with flammable gases and solvent vapors. Therefore, the maximum experimental safe gap cannot be increased with increased gap length. A maximum is reached, however, at a gap length of lg = 40–50-mm, which depends upon the type of dust and cannot be exceeded. This maximum is generally called the "quenching distance". It also follows the same correlation shown in Fig. 271. For gap sizes below the quenching distance but above the maximum experimental safe gap, there is an explosion halt or stop in the gap area, while there is also a re-ignition of the exterior mixture by the hot gases. For gap widths larger than the quenching distance, there will be an independent explosion propagation through the gap.

With this, Schuber confirmed the present knowledge, which originated from practical testing.

Rotary air locks will generally be safe against ignition propagation given the following criteria:

1) two vanes per side are engaged,
2) the vanes or the tips are made out of metal, and
3) the gap between rotor and housing is ≤0.2-mm.

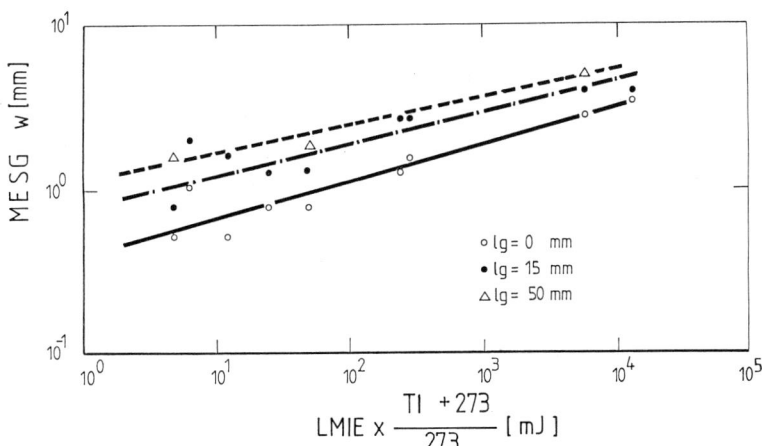

Fig. 271. Combustible dusts: Maximum experimental safe gap as a function of the product lowest minimum ignition energy (LMIE) times reduced ignition temperature TI

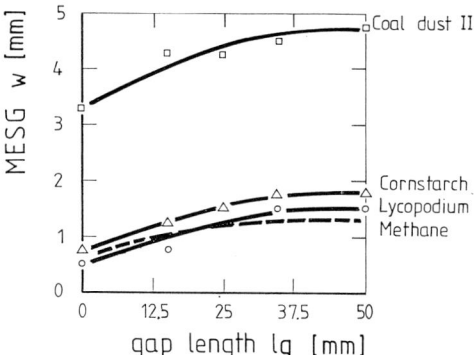

Fig. 272. Combustible dusts: Influence of the gap length lg upon the maximum experimental safe gap

This can be confirmed with information from Fig. 271. Sulfur, for example, has a very low minimum ignition energy (LMIE = 0.5 mJ) and also a low ignition temperature (TI = 250 °C).

The product of LMIE times reduced ignition temperatures gives ~1 mJ, which corresponds to a maximum experimental safe gap w = 0.4-mm referenced to a gap length lg = 0-mm (cross section of a circular gap with knife edges).

In case of an explosion, the rotary air lock has to be stopped immediately by a suitable detector in order not to pass burning or glowing product which may cause a second fire or explosion. Rotary air locks need not only be tested for their suitability as flame arresters but also must be tested for their pressure rating with appropriate explosion tests.

5.3.5.4 Rapid-Action Valves: Gate or Butterfly Type

Rapid action gate valves have the advantage that the closing device is normally outside the pipe cross section. The pipe area is completely open and can be built without pockets and dead corners, so that dust will not settle or accumulate. The valve body is made out of steel, cast steel, or stainless steel. The gate is made out of light, high tensile strength material so that short closing times will be possible. Special dampers have been developed in order to absorb the substantial shock forces from the closing device and to prevent the gate from springing back after closure.

At present, two types of rapid action gate valves can be recognized:

- **valves** with integral compressed air cylinder: In case of an explosion, an impulse of the central control unit ignites a detonator which breaks the holding pin of a catch at the prenotched location. The gate suddenly closes due to the pressure exerted on the driving piston (propellant pressure 4–6 bar) and is pneumatically dampened. The propellant may also be released by a fast-acting electromagnetic valve [134].
- **valves** connected to the compressed air cylinder via high-pressure hose. In case of an explosion, again an impulse from a central control unit will activate a deto-

nator which opens the valve of the compressed air cylinder. The propellant (pressure 10–40 bar), via a cylinder-piston system, closes the gate, which will be dampened through the plastic deformation of a braking device [135].

The functioning of both valve types can be tested in place. Rapid-action valves need an external operating device, which requires a suitable detector with control unit for triggering.

In case a vessel is protected by a design measure:

- explosion-resistant design for the maximum explosion pressure
- explosion-resistant design for the reduced maximum explosion pressure in conjunction with explosion pressure venting or explosion suppression,

an explosion can activate the downstream rapid-action valve through the following means:

- a rupture disc with a sensing device or
- the usual membrane pressure sensor with a corresponding low activation pressure.

If the activation is by an optical flame detector, which is mounted in the wall of the pipe, then the distance to the rapid-action valve must be in line with the anticipated flame velocity and the closing time of the valve. The rapid-action valve described above can be mounted in vertical, horizontal, or slanting pipelines.

Rapid-action valves have to be tested under explosion conditions to determine their effectiveness as a flame barrier and their pressure ratings before actual use in practice. Such a test arrangement is shown in Fig. 273.

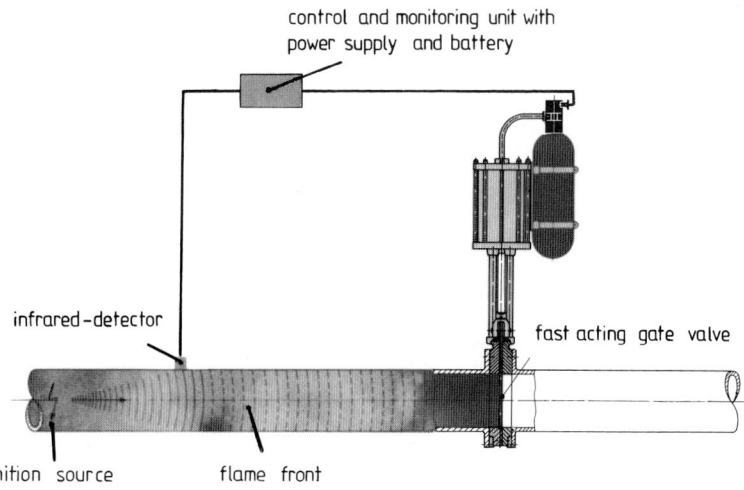

Fig. 273. Arrangement for the performance testing of rapid action valves with respect to flame containment and pressure shock resistance

234   5 Protective Measures Against the Occurrence and Effects of Dust Explosions

Before the start of the test, the pipeline up- and downstream from the valve is filled with an explosible fuel/air mixture. An optical flame detector senses the starting explosion, and its control unit activates the rapid-action valve, which stops the explosion analogous to practical applications. During such a test, the flame velocities and the explosion pressure (especially ahead of the valve) are recorded for the whole system.

The following are used as a test medium:

- very easily ignition-propagating combustible dusts (e.g., Lycopodium or aluminum fine dust)
- the flammable gas propane in order to cover the range of hybrid mixtures.

For the tests, the fuel concentration is systematically varied. The tests for flame propagation are carried out with low turbulent mixtures, the ones for pressure resistance with normal turbulent dust/air mixtures. Propane/air mixtures are basically ignited in a quiescent state. Such a test arrangement can also be modified by substituting an explosion vessel of, e.g., 2.4-m^3 (Fig. 274) for the pipeline ahead of the rapid-action valve.

The aim of the test is to subject the valve to an explosion pressure of approximately 10 bar gage without allowing a flame propagation.

Figure 275 shows an explosion protection rapid-action valve which fulfills these requirements. It was developed from a standard gate valve for dust service through systematic tests [136]. It is manufactured in the nominal sizes DN 50–650 and, depending upon diameter, has closing times between 10 ms (DN 50) and 50 ms (DN 650). It has already been pointed out that the end-flanges of longer pipelines will be exposed to markedly higher explosion pressures, especially in the case of explosions with dusts of the dust explosion class St 3 (aluminum) or hybrid mixtures

Fig. 274. Test arrangement with an explosion vessel for the testing of rapid-action valves for flame propagation and explosion pressure shock resistance

5.3.5 Technical Diversion or Arresting of Explosions

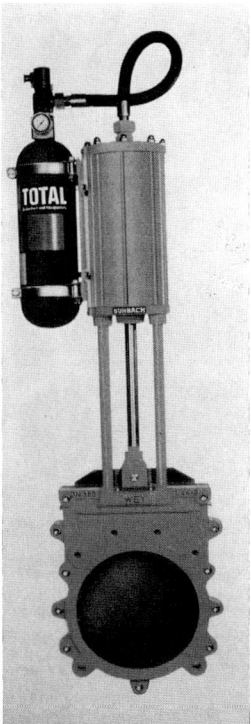

Fig. 275. Explosion protection valve WEY

(Fig. 137). The reason for this is not the development of a detonation, but instead the acceleration of the explosion flame into the precompressed mixture ahead of the closed pipe. These amplified effects of an explosion also occur at the location of a closed rapid-action valve if it is installed in a pipeline more than 10-m away from a vessel which is protected by design measures. The result is a heavily deformed plate (Fig. 276) and naturally propagating ignition of the fuel/air mixture.

As a remedy, a special venting device was designed (Fig. 277), at first for a pipeline DN 400. It entails a 3-m-long pipe section with four equally spaced flanged venting slots around the pipe circumference. The total vent area corresponds to 4.5 times the cross section of the pipe. The vent openings are closed with lightweight aluminum plates supported in rubber moldings. Two steel slings prevent the plates from flying away in case of an explosion.

The venting device is mounted as close as possible to the rapid-action valve in order to accomplish a decrease in the loading pressure of the valve gate (Fig. 278). Systematic explosion tests were made to determine the effectiveness of such a venting device in conjunction with an activated explosion valve DN 400. They showed that the activation pressure of the venting device is of the utmost importance for the reduction of the explosion pressure at the location of the device. The lower the pressure is, the more unburnt mixture will be exhausted through the vent

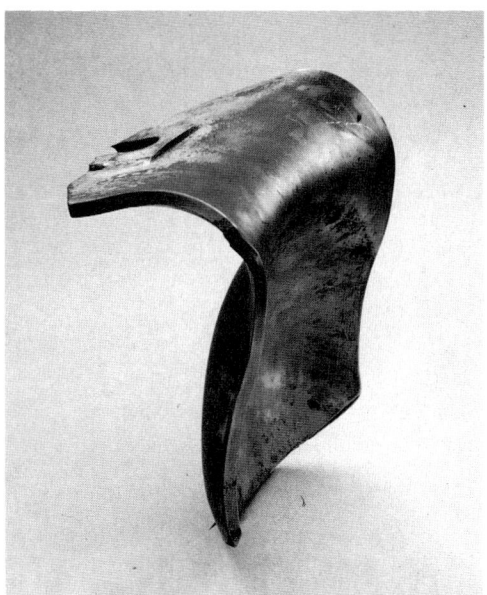

Fig. 276. Deformed valve-plate due to exposure to excessive explosion pressure

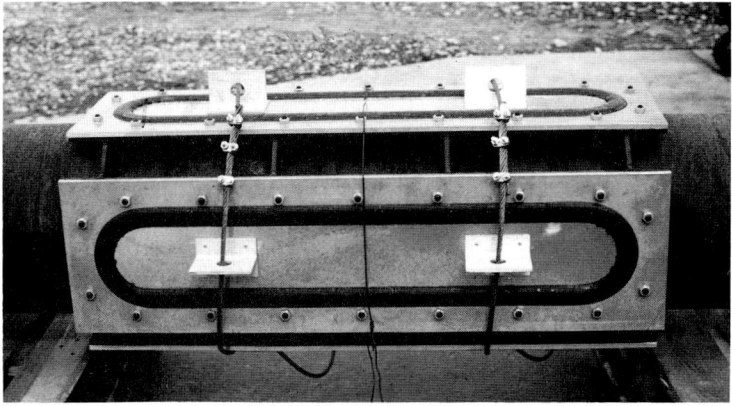

Fig 277. Venting device for the explosion protection valve WEY

ahead of the flame front, which will not participate in the combustion. The activation pressure has to be kept positively below $p_a \leq 0.5$ bar in order to maintain the pressure loading in the gate area below the design pressure of the valve.

Figure 279 shows the influence of an extremely high and low activation pressure upon the pressure development and flame velocity of an aluminum-dust explosion.

5.3.5 Technical Diversion or Arresting of Explosions

Fig. 278. Explosion protection valve WEY with upstream venting device

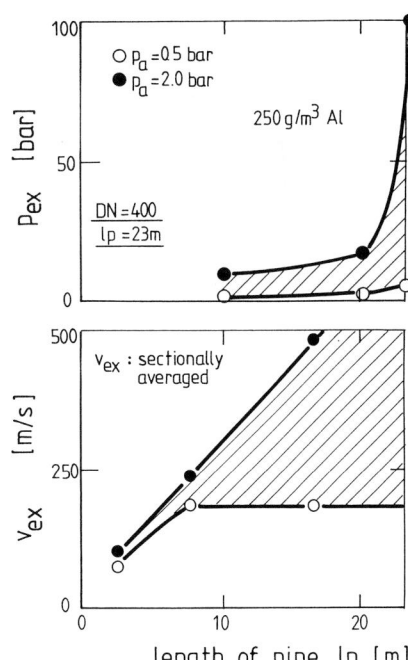

Fig. 279. Influence of the activation pressure of the venting device upon the flame velocity and pressure development of an aluminum-dust explosion (explosion valve installed at 23-m mark)

Now, the question can be raised whether the arrangement of an additional venting device in the pipeline will allow the use of an explosion valve with a nominal 10 bar rating in case of a detonation. Fine aluminum dust at a concentration of 375 g/m^3 has a detonation-like behavior in a 33-m-long pipeline DN 400 when attached to a 2.4-m^3 ignition vessel (v_{max} = 3000 m/s, p_{max} = 37 bar). Such a detonation can be mitigated if, at first without an explosion valve, a venting device having an activation pressure of $p_a \leq 0.5$ bar (Fig. 277) is installed at the 0-m mark (Figs. 280 and 281), plus another at the 10- or the 20-m mark of the pipeline. A venting device installed at the 10-m mark (Fig. 282) has the most favorable relief effectiveness.

Both the explosion pressure and the flame velocity decrease towards the end of the pipe, the anticipated location of the explosion valve.

If at the end of the pipe an additional explosion valve with a venting device is installed, then the behavior of the pressure and flame inside the pipe cannot be distinguished from the one obtained with a pressure-resistant closure at the pipe end (Fig. 283). The valve acted perfectly as a flame barrier, but the gate was deformed by the elevated reflected pressure. From this, the conclusion has to be that the re-

Fig. 280. Venting device installed at the 0-m mark

Fig. 281. Arrangement as per Fig. 280 with 375 g/m^3 fine aluminum dust

5.3.5 Technical Diversion or Arresting of Explosions

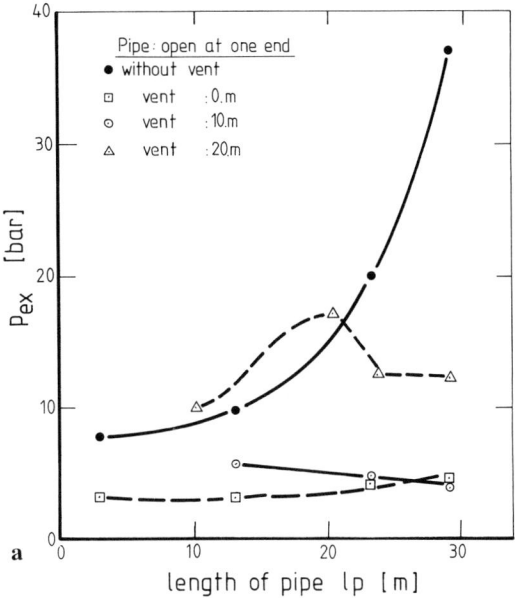

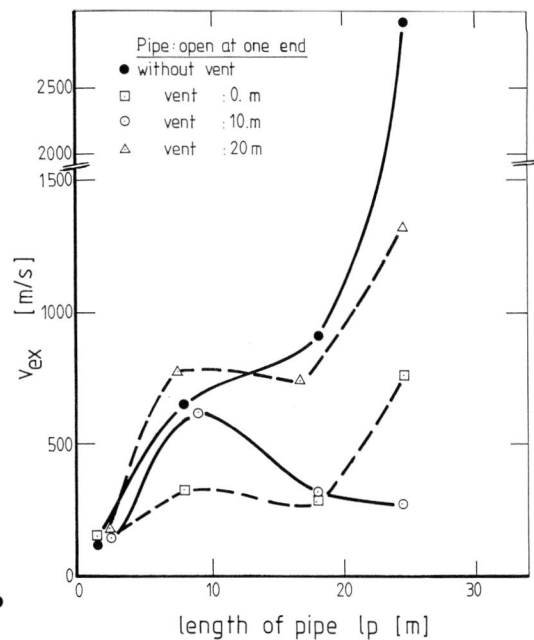

Fig. 282 a/b. Influence of the location of the venting device upon the explosion characteristics of fine aluminum dust (375 g/m^3) (2.4-m^3/DN 400, lp =33-m, without explosion valve)

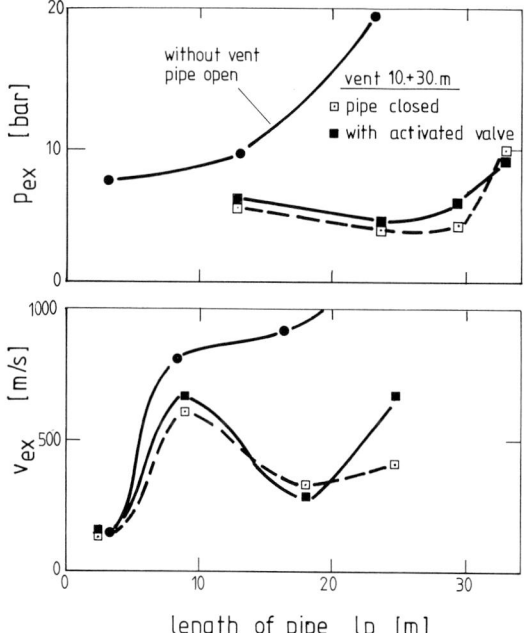

Fig. 283. Influence of the test arrangement upon the explosion characteristics of fine aluminum dust (375 g/m³) (pipeline DN 400, lp = 33-m)

quired pressure rating of a rapid-action valve must be explosion-resistant for a nominal 10 bar pressure and not explosion pressure shock-resistant. It will then remain functional after being exposed to a detonation-like behavior of combustible dusts, provided that a venting device is installed ahead of the valve and at the 10-m mark. For longer pipelines, such a venting device has to be installed every 20-m. Due to the development of flames and pressure, such an installation is limited to an open air application.

Figure 284 shows a closed explosion valve with an activated venting device.

The test results with an explosion valve show that the problems inherent in the threat of a violent dust explosion can be mastered and that economical solutions are possible (no excessive explosion rating for the valve). The stoppage of a dust explosion in a pipeline may also be accomplished by rapid-acting butterfly valves which are controlled by detectors.

The make shown in Fig. 285 is pneumatically closed by the escaping air from a compressed air cylinder which again has a valve actuated by a detonator. The requirements for such a system are the same as for rapid-action gate valves.

Fig. 284. Closed explosion valve with activated venting device after a detonation test

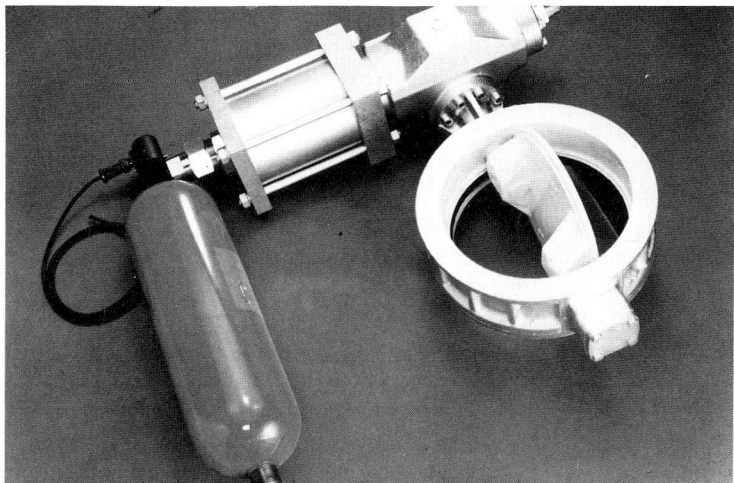

Fig. 285. Rapid-action butterfly valve

5.3.5.5 Rapid-Action Valve: Float Type

Often it is necessary to prevent an explosion from leaving a piece of equipment, e.g., a grinding installation which is protected through design measures, and then entering the vent system which has no pressure resistance. One possible solution for this is the float-type, rapid-action valve (Ventex), which is activated by the ex-

Fig. 286. Explosion pressure-activated rapid-action float valve (Ventex ESI)

plosion pressure or a detector. For such a valve type ESI (Fig. 286), the conveyed medium flows through the valve. The inner part of the valve consists of a float which is supported by bushings and therefore can move axially in either direction. The center position is controlled by springs. The spring is generally loaded for a maximum flow velocity v = 24 m/s, based on the cross section of the pipe [128].

In case of an explosion, the valve closes automatically because of the explosion pressure which precedes the flame front. For the valve to function, the explosion velocity has to be >24 m/s or the pressure differential before and after the float ≈0.1 bar. Upon closing, the float is pressed against a gasketed seat and locked in place by a holding mechanism. Therefore, the valve remains closed after activation, but can be brought back into the normal center position through an externally operated release mechanism. The closed valve position can be optically indicated by electrical limit switches. The Ventex valve functions axially in either direction. Such a valve can only be installed in a horizontal pipeline and is in general only suitable for streams having a relatively small dust loading (e.g., clean air duct downstream from filters). The closing time of the valve is in the range of 50 ms. Therefore, its location should be approximately 5-m away from the vessel which is protected by design measures. In case the safety measure "explosion pressure venting" is used, its static venting pressure has to be assured at a higher level than the required differential pressure for the valve closure. As already mentioned, a certain explosion pressure is needed to initiate the closing action of the valve. If such a pressure is not reached, the flame will pass through and the explosion will propagate beyond the valve location. If such a range of uncertainty cannot be tolerated, then the valve can also be actuated by external means. Such actuation can be effected through pressure impingement of the valve (Fig. 287) from a cylinder axially through a hemispherical nozzle (Fig. 239). The cylinder contains nitrogen at a certain pressure and the nozzle is the same type usually used for suppression. The detonator-activated valve of the cylinder requires an optical flame detector towards the source of explosion.

For the new Ventex valve P, the external energy which is required for its closure is continuously supplied from a 6 bar compressed-air system. A suitable detector

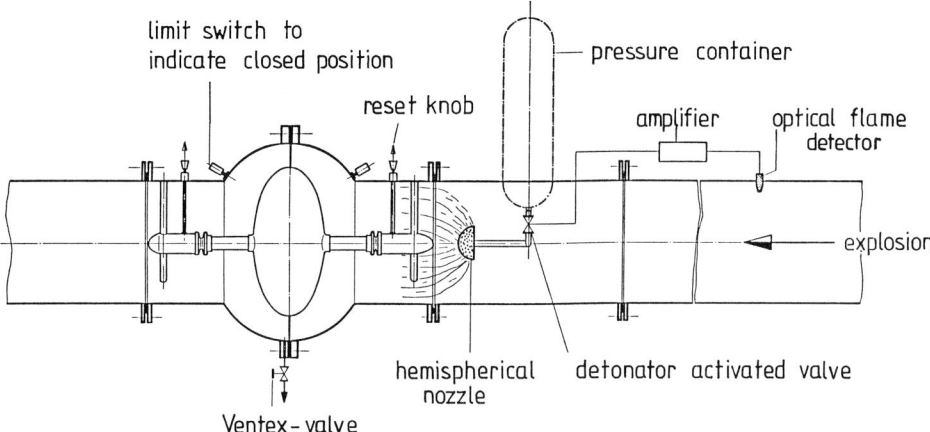

Fig. 287. Ventex valve ESI with externally assisted activation (schematic)

triggers, via a control unit, the magnetic lock of the compressed air cylinder. Such a detector may be a rupture disc with signaling device or a membrane pressure sensor on the protected vessel. For a pipeline, it is an optical flame sensor installed in the pipe wall. Once the closing action is started, both externally activated Ventex-type valves will be maintained in a closed position. The re-setting back to operating position is done manually.

The externally activated Ventex valves work only in one direction and are only suited for a horizontal installation in pipelines [128].

Externally activated Ventex valves have the advantage that they remain functional (the uncertain pressure range excluded) even if the triggering detector should fail.

5.3.5.6 Explosion Diverter

A diverter (Fig. 288) which turns the flow 180° is especially economical for disengaging explosions.

It consists of pipelines which are interconnected by a special pipe section. The section is closed from the atmosphere by a cover or rupture disc. Parts which may fly away must be restrained by a cage. The main thrust of an explosion starting in a longer pipeline ahead of the equipment needing protection will be diverted upwards after the coverplate has lifted. Any restarted explosion in the downward pipe, if it occurs at all, will enter the equipment protected by, e.g., design measures with a much lower explosion velocity and pressure. Flame jet ignition with pressure piling (Fig. 248) will be prevented with such an arrangement. Only explosion barriers or fast-acting valves will prevent the explosion transfer entirely if the downstream equipment is only designed for atmospheric exposure, i.e., preventive explosion protection is thereby practiced. In conclusion, Fig. 289 shows the activated explosion diverter installed in a pipeline connecting a silo and a filter during a corn starch dust explosion.

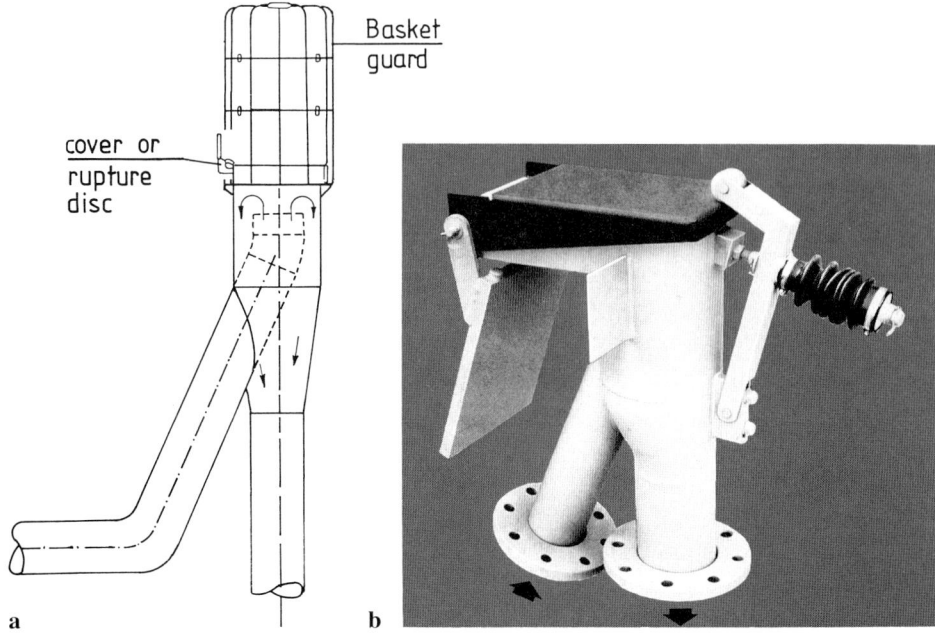

Fig. 288 a/b. Explosion diverter (**a**) schematic presentation, **b**) actual diverter) [137]

Fig. 289. Corn starch explosion in a 20-m^3 silo with activated explosion diverter

5.3.6 Conclusions

In case the preventive explosion protection approach cannot be used for installations which manufacture, process, and convey combustible dust, a number of "design measures" can be used for their protection. Such measures will not prevent a dust explosion but will limit the effects to an unobjectionable level. These measures basically require a certain explosion resistance of the vessels, equipment, and pipelines, which may be built explosion pressure-resistant or explosion pressure shock-resistant. Whereas explosion pressure-resistant vessels and equipment designed for the full maximum explosion pressure or a reduced maximum explosion pressure have proven themselves for decades, the pressure shock-resistant design is considered safe based on today's limited knowledge.

Vessels, equipment, and pipelines can be designed for the full maximum explosion pressure. This is generally feasible for sizes up to a few m^3. If the threat of an explosion comes from non-toxic combustible dust which will not damage the surroundings, then explosion pressure venting is a viable, very effective design measure. In addition, within limits, certain reductions in relief area are possible in comparison with the valid guidelines, provided that the maximum explosion pressure of the dust does not exceed 9 bar. Such a protective measure can also be used in pipelines and for silos by incorporating the newer findings.

The effectiveness of explosion suppression systems has been substantially improved so that this proven design measure can be successfully used for large and larger vessels independent of location. Explosion diversion is a flanking measure for the above-mentioned design measures that is helpful in separating equipment from the systems which use preventive explosion protection.

Extensive practice-related tests in large pipelines, bucket elevators, and silos have shown that extinguishing barriers can be used without reservation to halt or stop an explosion in practice.

The same is true for rotary air locks which have been tested for ignition breakthrough and pressure resistance. Early results are now available pertaining to flame propagation of dust/air mixtures through narrow gaps, e.g., rotor and housing of an air lock.

The findings pertaining to the use of rapid-action valves for the restriction of the course of dust explosions have been substantially expanded. Such valves may now be successfully used in practice, even with the danger of an extreme course of an explosion (pressure piling, detonation).

Rapid-action valves have been used as flame barriers against dust explosions for a long time in practice, with a proven record.

Explosion diverters are economical solutions to disengage explosions. They find more and more use in practical installations, not only for limiting the course of an explosion, but also to prevent flame jet ignition (maybe at a possible precompression) in vessels or equipment which is protected through design measures.

All in all, there is no problem in explosion technology which cannot be solved with the above-mentioned safety measures in conjunction with an explosion diverter.

6 Concluding Remarks

In the years following the first two editions of my book "Explosions: Course, Prevention, Protection" in 1978/1981, many problems pertaining to the explosion technology of combustible dusts have been studied and solved. This related not only to the improvements of the test methods needed to detect the dangers inherent in dust layers and airborne dust, but also the explosion protection through preventive and design means. Open gaps were closed through methodical research.

Therefore, I did not honor the request of my publisher to print a third edition of my book, but decided to undertake a complete revision. Such a revision, however, is limited to combustible dusts, in line with the revised VDI guideline 2263: "Dust Fires and Dust Explosions: Hazards-Assessment-Protective Measures", which has been just issued at the time this book was published.

This book again attempts to survey the present state of the art and starts with a historical review of the development of the problematic nature of "dust explosions". It is shown that such findings are not of recent date but had already been known for approximately 200 years.

All organic and metallic dusts are capable of exploding, but not all have to be considered explosible while being manufactured, processed, and conveyed in practice. The new findings pertaining to the ignition capabilities of electrically and mechanically generated sparks, electrostatic discharges, and hot surfaces in dust/air mixtures facilitate the use of the preventive safety measure "prevention of effective ignition sources". However, its practical application assumes the knowledge of the safety characteristics of the fuel plus expert judgement. The presence of an explosible dust/air mixture in part of or in the whole of an installation does not always present an explosion danger.

Substantial improvements have been made in the field of explosion protection through design measures and through the flanking explosion diversion measures.

The book was written for the practical user and abstains from theoretical dissertations, which are rare in the field of combustible dusts. The numerous figures were not only chosen to document for the reader the danger of a dust explosion but also to visually present solutions for meeting such a danger.

The reader should realize: Dust explosions can be mastered!

7 Acknowledgements

I thank, without exception, all members of the national and international professional organizations and divisions for the numerous, useful discussions with which they helped to create the image of the safety technology available to control the dangers of dust explosions. This is especially true for Mr. C. Donat (Farbwerke Hoechst AG), who retired in the meantime.

I extend my thanks to the numerous equipment manufacturers for their unselfish support for the very elaborate and costly tests.

Included in my thanks are the "Bundesministerium für Forschung und Technologie", the "Berufsgenossenschaft der chemischen Industrie" and the "Berufsgenossenschaft Nahrungsmittel und Gaststätten", for their financial help in carrying out various test programs.

I owe special thanks to Mr. A. Schaerli, the previous head of the central safety department of Ciba-Geigy, Ltd. for the appreciation he always showed for the work done in the explosion technology group, which formerly reported to him. He exerted considerably energy and persevered for the improvement of the test conditions and the expansion of the facilities in the laboratory and the open air station in the Jura, Switzerland. Without his support, a major portion of the tests mentioned in this book, especially the large-scale tests, would not have been possible.

I thank also Mr. G. Zwahlen, the head of the safety test station at Ciba-Geigy, Ltd. for his contribution and always helpful collaboration.

Included in my thanks are all my former and present colleagues. Their special effort was decisive for the success of the tests, and their results are used as the basis of this book.

8 Appendix

8.1 Explosion Pressure Venting

8.1.1 Vessel: Area Determination by Calculation or Nomogram

Recently, Radandt developed a model incorporating the newest findings [138] for the correlation of vent area A with the vessel size V of a cubic vessel to be protected against dust explosions of the dust explosion classes St 1 and St 2 with a maximum explosion pressure $p_{max} = 9$ bar.
It has the general form:

$$A = \left(a + \frac{b}{p_{red,max}}\right) \times V^c$$

The values of the factors can be taken from Table 24.
The above figures are valid for a venting system consisting of rupture panels having a static venting pressure $p_{stat} = 0.1$ bar.

Table 24. Factors for the calculation of the vent area A as a function of vessel size V ($p_{max} = 9$ bar, $p_{stat} = 0.1$ bar, $p_{red,max} \leq 2$ bar)

Dust explosion class	$p_{red,max}$ [bar]	a	b	c
St 1	<0.5	0.04	0.021	0.741
	≥0.5	0.04	0.021	0.766
St 2	<0.5	0.048	0.039	0.686
	≥0.5	0.048	0.039	0.722

Analogous to Fig. 190, the required vent area for dusts belonging to the dust explosion classes St 1 and St 2 can be taken from the newly developed nomograms (Fig. 290/291) for a given vessel size V and a known pressure ratio (equivalent to the reduced maximum explosion pressure $p_{red,max}$) [139].

8.1.2 Elongated Vessels (Silos)   249

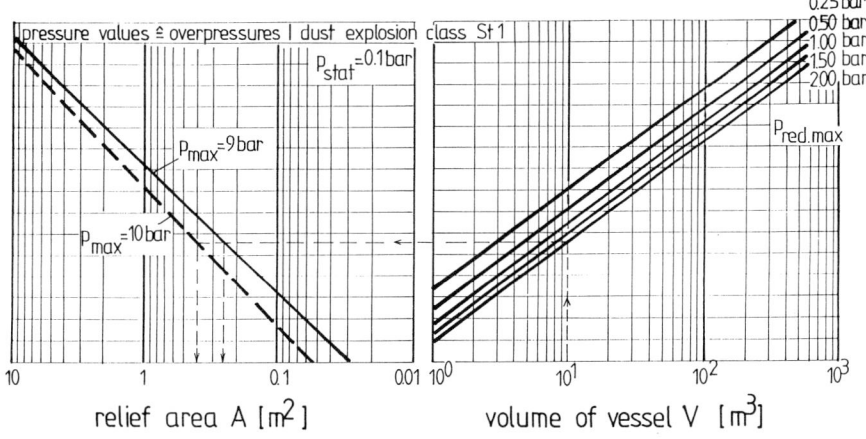

Fig. 290. Nomogram based on dust explosion class St 1 ($p_{stat} = 0.1$ bar)

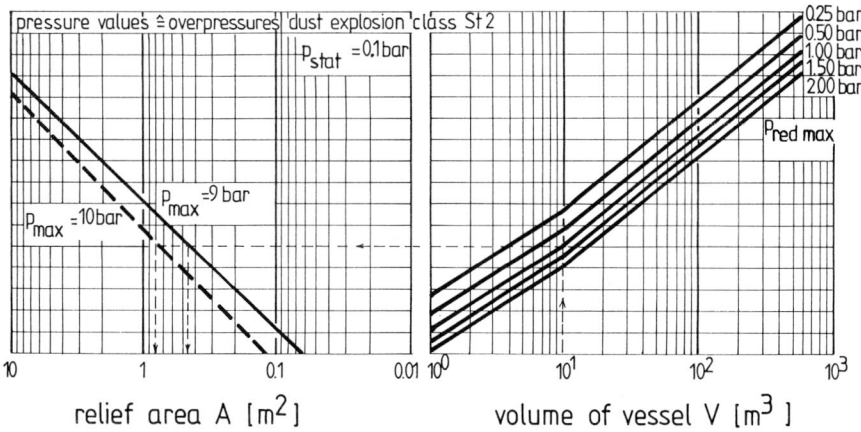

Fig. 291. Nomogram based on dust explosion class St 2 ($p_{stat} = 0.1$ bar)

8.1.2 Elongated Vessels (Silos)

In pneumatically conveying combustible dusts directly into a pressure-vented 20-m^3 silo [140], the optimum material-to-air ratio which results in the reduced maximum explosion characteristics markedly depends upon the product density (Fig. 292). This is also true for cubic vessels. In order to prepare a dust/air mixture of optimum concentration, more product is needed for heavier dusts than for lighter ones, which is obvious.

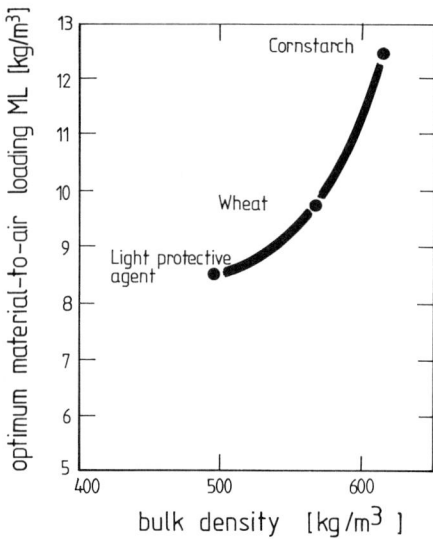

Fig. 292. Correlation of density with optimum material-to-air ratio for products directly conveyed into a 20-m³ silo

There is a linear correlation of air volume Q at the optimum material-to-air ratio with the explosion characteristics independent of the dust type. This is shown in Fig. 293 for a constant vent area A = 0.5-m². The pressure-area curve especially shows a clearly proportional behavior with the dust-specific K_{St}-value (Fig. 294).

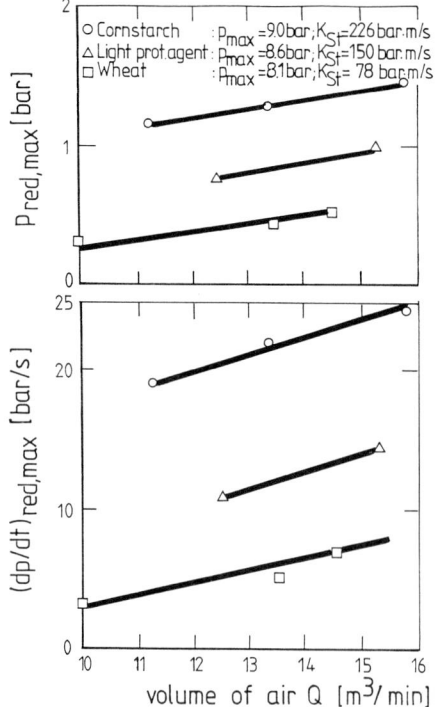

Fig. 293. Correlation of air volume Q with the reduced maximum explosion characteristics V = 20-m³ silo; A = 0.5-m²; optimum material-to-air ratio

8.1.2 Elongated Vessels (Silos)

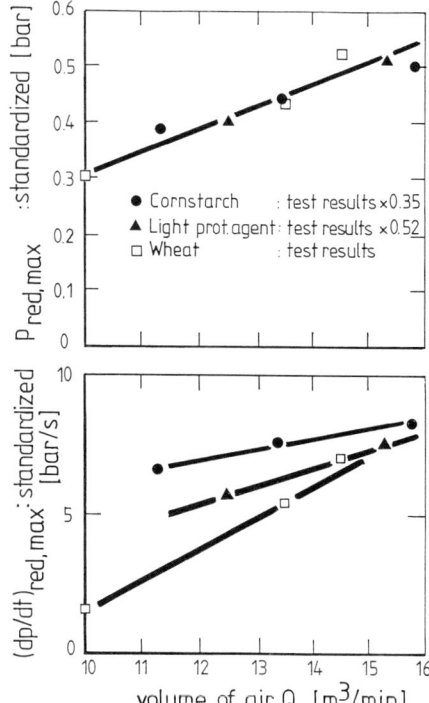

Fig. 294. Correlation of air volume Q with the standardized reduced maximum explosion characteristics V = 20-m³ silo; A = 0.5-m²; optimum material-to-air ratio

Such an observation allows a statement with regard to the correlation of vent area A with the reduced maximum explosion pressure $p_{red,max}$ for a 20-m³ silo and dusts reacting with varying violence (Table 25, Fig. 295).

Table 25. Direct pneumatic conveying: Correlation of vent area A with reduced maximum explosion pressure $p_{red,max}$ for different K_{St}-values
(V = 20-m³ silo; H/D = 6.25; $p_{max} \leq 9$ bar; $p_{stat} = 0.1$ bar; Q = 15–16 m³/min; optimum material-to-air ratio)

K_{St} [bar · m/s]	50	100	200	300
A [m²]		$p_{red,max}$ [bar]		
0.2	0.82	1.43	2.67	3.93
0.5	0.46	0.67	1.29	1.91
1.0	0.23	0.38	0.68	0.98
1.5	0.14	0.24	0.48	0.70
2.0*	0.10	0.20	0.38	0.56

* Silo top completely used as vent area

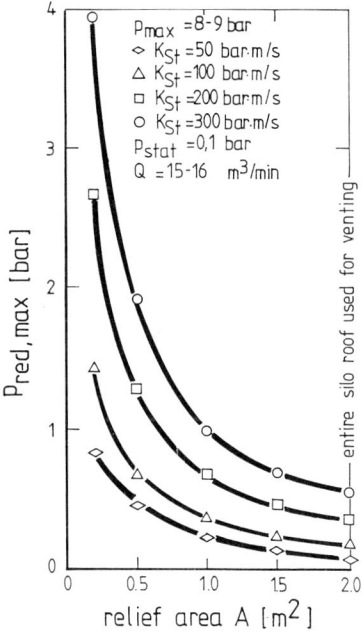

Fig. 295. Pressure-area curves of combustible dusts pneumatically conveyed directly into a 20-m³ silo (optimum material-to-air ratio)

Again it is emphasized that the figures in Table 25 are only valid for air volumes $Q \leq 16$ m³/min (see Fig. 227) and for an optimum material-to-air ratio (see Fig. 292). Any deviation from such a ratio lowers the reduced maximum explosion pressure.

In addition, Fig. 295 indicates that with the full use of the silo top as a vent area ($A = 2.0$-m²), only the pressure generated by weakly reacting dusts ($K_{St} \leq 50$ bar·m/s) will be equivalent to the static venting pressure of the relief device. For more violent reacting dusts, however, the pressure will be higher, requiring a higher explosion pressure rating so that the silo will not be destroyed in case of explosion.

In adhering to the model developement for cubic vessels, Radandt [138] gives additional mathematical correlations for the sizing of vent areas A for larger silos ($V > 20$-m³). The factors for the equation shown in sect. 8.1.1 can be taken from Table 26.

Table 26. Factors for the sizing of vent areas A as a function of silo size V ($p_{max} = 9$ bar, $p_{stat} = 0.1$ bar)

Dust explosion class	$p_{red,max}$ [bar]	a	b	c
St 1	≤ 2	0.011	0.069	0.776
St 2	≤ 2	0.012	0.114	0.720

8.1.2 Elongated Vessels (Silos)

The calculated values from the model giving the correlation of reduced maximum explosion pressure with vent area A coincide nicely with the actual measured values for the 20-m³ silo (height/diameter ratio = 6.25) (Fig. 295). Based on the correlation found for cubic vessels, larger volumes will require more vent area than expected from the cubic law.

In many cases, the venting of a silo is limited by the available top area (see sect. 5.3.3.3).

For a given volume, but increasing height/diameter ratio, such a top area becomes smaller, and, therefore, the required explosion pressure $p_{red,max}$, which can either be interpolated from Table 25 or calculated from the transformed equation in sect. 8.1.1.

$$p_{red,max} = \frac{b \cdot V^c}{A - a \cdot V^c}$$

The factors are to be taken from Table 26, and A is the available area at the top of the silo.

9 References

1. BIA: Dokumentation Staubexplosionen (Analyse und Einzeldarstellung), Berufsgenossenschaftliches Institut für Arbeitssicherheit – Report 4/82
2. W. Bartknecht: Staubexplosionen – Ein historischer Überblick, VDI-Berichte 494, Seite 11–24, 1984
3. W.H. Geck: Zündfähige Industriestäube, Verlag des Vereins Deutscher Ingenieure, 1954
4. P. Beyersdorfer: Staubexplosionen, Verlag von Theodor Steinhoff, Dresden und Leipzig, 1925
5. L. Darmstaedter: Handbuch der Geschichte der Naturwissenschaften und Technik, Verlag von J. Springer, 1908
6. Berggewerkschaftliche Versuchsstrecke: 75 Jahre Berggewerkschaftliche Versuchsstrecke, Festschrift 1968
7. G. Pellmont: Explosions- und Zündverhalten von hybriden Gemischen aus brennbaren Stäuben und Brenngasen, Diss. ETH Zürich Nr. 6498, 1979
8. Chemische-Technische Reichsanstalt: Jahresbericht 1933–1935 S. 372–388
9. W. Gliwitzky: Über Staubexplosionen, ihre Grundlagen und Verfahren zu ihrer Untersuchung, Elektrotechnische Zeitschrift, 59. Jahrgang, Heft 43, 1938
10. H. Selle: Die chemischen und physikalischen Grundlagen der Verbrennungsvorgänge von Stäuben, VDI-Berichte, Band 19, 1957
11. G. Leuschke: Licht- und Wärmestrahlung, Handbuch der Raumexplosionen v. H. Freytag, Verlag Chemie GmbH, Weinheim/Bergsstrasse, 1965
13. B. Howard, H. Essenhigh: Pyrolyses of Coal Particles in Pulverised Fuel Flames, Industrial and Engineering Chemistry, Process, Design and Development, Vol 6. Nr. 1, S. 74–84, 1967
14. H. Essenhigh, B. Howard: Towards an Unified Combustion Theory, Industrial and Engineering Chemistry, Vol 58, Nr. 1, S. 15–23, 1966
15. Vital: Annales des mines, Ser. 7 Vol. 20, S. 180, 1875
16. H. Selle: Die Grundzüge der Experimentierverfahren zur Beurteilung brennbarer Industriestäube, VDI-Berichte, Band 19, 1957
17. VDI-Kommission Reinhaltung der Luft: Internationaler Erfahrungsaustausch: "Sichere Handhabung brennbarer Stäube" am 20. und 21.6.1979, Düsseldorf, S. 25 (Ab Manuskript gedruckt: 1981)
18. B. Mason, S. Taylor: Ind. Engng. Chem. 29, S. 626, 1937
19. H. Selle, J. Zehr: Experimentaluntersuchungen von Staubverbrennungsvorgängen und ihre Betrachtung vom reaktionsthermodynamischen Standpunkt, VDI-Berichte, Band 19, 1957
20. J. Lütolf: Untersuchungen von Stäuben auf Brand- und Explosionsgefahr, VDI-Berichte, Band 165, 1971
21. W. Bartknecht: Brenngas- und Staubexplosionen, Forschungsbericht F 45 des Bundesinstitutes für Arbeitsschutz, 1971
22. J. Zehr: Die physikalische Kennzeichnung der Stoffeigenschaften, VDI-Berichte, Band 19, 1957
23. VDI-Kommission Reinhaltung der Luft: Staubbrände und Staubexplosionen: Gefahren – Beurteilung – Schutzmassnahmen, VDI-Richtlinie 2263, 1986, Beuth-Verlag GmbH, Berlin und Köln
24. VDI-Fachausschuss "Brennbare Stäube": Verhütung von Staubbränden und Staubex-

plosionen, VDI-Richtlinie 2263, Ausgabe August 1969, Beuth-Verlag GmbH, Berlin und Köln
25. R. Meldau: Handbuch der Staubtechnik, Band 1, Grundlagen, Düsseldorf 1952
26. Berthold/Löffler: Lexikon sicherheitstechnischer Begriffe in der Chemie, Verlag Chemie, Weinheim, 1981
27. G. Leuschke, R. Osswald: Bedeutung und Ermittlung von sicherheitstechnischen Kengrössen brennbarer Stäube, VDI-Berichte 304, S. 29–38, 1978
28. J. Lütolf: Kurzmethoden zur Prüfung brennbarer Stäube, VDI-Berichte 304, S. 39–46, 1978
29. Expertenkommission für Sicherheit in der chemischen Industrie der Schweiz: Heft 1: Sicherheitstests für Chemikalien, S. 6, 1985
30. (anonym): Amtsblatt der Europäischen Gemeinschaften Nr. L 251, 27, (19.9.1984), S. 63. Prüfvorschrift A 10: Entzündlichkeit – Feste Stoffe
31. L. Bretherick: Handbook of Reactive Chemical Hazards, S. 60–63, Verlag Butterwords, London 1979
32. Expertenkommission für Sicherheit in der chemischen Industrie der Schweiz: Heft 1: Sicherheitstests für Chemikalien, S. 7, 1985
33. Th. Grewer, H. Schacke: Oxidations- und Zersetzungsreaktionen in Staubschüttungen, VDI-Berichte 494, S. 145–149, 1984
34. (anonym): DIN 57 165, Sept. 1983
35. H. Freytag: Handbuch der Raumexplosionen, Verlag Chemie GmbH, Weinheim/Bergstrasse, 1965
36. (anonym): DIN IEC 31H (Co)3, November 1982, zugleich: VDE 0170/0171, Teil 102
37. Th. Grewer: Zur Selbstentzündung von abgelagertem Staub, Staum, Reinhaltung in der Luft, 31, Nr. 3, S. 97, 1971
38. (anonym): Amtsblatt der Europäischen Gemeinschaften Nr. L 251 27 (19.9.1984), S. 86, Prüfvorschrift A 16: Selbstentzündlichkeit – Feste Stoffe
39. J. Lütolf: Untersuchungen von Stäuben auf Brand- und Explosionsgefahr, Staub, Reinhaltung der Luft, 31, Nr. 3, S. 93, 1971
40. N. Semenow: Chemical Kinetics and Chain Reactions, The Clarendon Press, Oxford 1935
41. D. A. Frank-Kamenetzkii: Stoff- und Wärmeübertragungen in der Chemischen Kinetik, Springer-Verlag Berlin, Göttingen, Heidelberg, 1959
42. R. Gygax: Risiken unerwünschter Selbstwärmevorgänge in ungerührten Behältern. Swiss Chem 5, Nr. 9a, S. 31, 1983
43. W. Berthold, M. Heckle, H.J. Lüdecke, A. Ziegner: Einfache Untersuchungsmethoden zur Vermeidung von Wärmeexplosionen in Grossbehältern, Chemie-Ing.-Technik 47, S. 368, 1957
44. Th. Grewer: Unerwünschte Reaktionen organischer Stoffe als Gefahren in Chemieanlagen, Chemie-Ing.-Technik 51, S. 928, 1979
45. (anonym): Amtsblatt der Europäischen Gemeinschaften Nr. L 251, 27, (19.9.1984), S. 74, Prüfvorschrift A 14: Explosionsgefahr
46. H. Koenen, K. H. Ide, K. H. Swart: Sicherheitstechnische Kenndaten, explosionsfähiger Stoffe, Explosivstoffe, 9, S. 4–13, 30–42, 195–197, 1961
47. H. Koenen, K. H. Ide, W. Haupt: Über die Prüfung explosiver Stoffe. IV. Ermittlung der Schlagempfindlichkeit explosiver Stoffe von fester, flüssiger und gelatöser Beschaffenheit, Explosivstoffe, 6, S. 178–189, 202–214, 223–235, 1958
48. H. Koenen, K. H. Ide: Über die Prüfung explosiver Stoffe: I. Ermittlung der Reibeempfindlichkeit, Explosivstoffe, 3, S. 57–65, 89–93, 1955
49. H. Koenen, K. H. Ide: Über die Prüfung explosiver Stoffe: III. Ermittlung der Empfindlichkeit explosiver Stoffe gegen thermische Beanspruchung in einer Erhitzungskammer mit verschiedenen definierten Öffnungen (Stahlhülsenverfahren), Explosionsstoffe, 4, S. 119–125, 143–148, 1956
50. A. Eucken: Grundriss der Physikalischen Chemie, 6. Auflage, Leipzig, 1944
51. W. Jost: Explosions- und Verbrennungsvorgänge in Gasen, Verlag Julius Springer, Berlin 1939
52. DIN 20 163: Begriffe und Benennungen der Sprengtechnik

53. J. Zehr: Anleitung zu den Berechnungen über die Zündgrenzwerte und die maximale Explosionsdrücke, VDI-Berichte Nr. 19, S. 63–68, Düsseldorf: VDI-Verlag 1957
54. J. Schönwald: Vereinfachte Methode zur Berechnung der unteren Explosionsgrenze von Staub/Luft-Gemischen, Staub-Reinhaltung der Luft 31, Heft 9, 1971
55. VDI-Kommission Reinhaltung der Luft: Druckentlastung von Staubexplosionen, VDI-Richtlinie 3673, Absatz 12, Juni 1979, Beuth-Verlag GmbH, Berlin und Köln
56. International Standard ISO 6184/I: Explosion Protection Systems – Part I: Determination of Explosion Indices of Combustible Dusts in Air, International Organization for Standardization, 1985
57. R. Siwek: Bericht über Versuche zur Entwicklung einer Laborapparatur für die Bestimmung der Explosionskenngrössen brennbarer Stäube, HTL-Winterthur, Semesterarbeit, 1976
58. R. Siwek: 20 l-Laborapparatur für die Bestimmung der Explosionskenngrössen brennbarer Stäube, HTL-Winterthur, Diplomarbeit, 1977
59. R. Siwek: Experimental Methods for the Determination of Explosion Characteristics of Combustible Dusts, 3rd Int. Symposium on Loss Prevention and Safety Promotion in the Process Industries, Basel, Vol. 2, p. 839–850, 1980
60. Th Glarner: Temperatureinfluss auf das Explosions- und Zündverhalten brennbarer Stäube, Diss. ETH Zürich Nr. 7350, 1983
61. BIA: Brenn- und Explosionskenngrössen von Stäuben, Forschungsbericht Staubexplosionen des Berufsgenossenschaftlichen Institutes für Arbeitssicherheit, März 1980
62. E. Schenk: Explosions- und Zündverhalten von Flockmaterial, VDI-Berichte 494, S. 53–57, 1984
63. W. Bartknecht: Untersuchung des Explosions- und Zündverhaltens brennbarer Stäube und hybrider Gemische, Schriftenreihe: "Humanisierung des Arbeitslebens", Band 64, Herausgeber: Der Bundesminister für Forschung und Technologie, VDI-Verlag, Düsseldorf, 1985
64. E. W. Scholl: Löschmittelsperre zur Verhinderung der Fortpflanzung von Schlagwetter- und Kohlenstaubexplosionen, Dissertation TH Clausthal, 1968
65. N. Kalkert: Theoretische und experimentelle Untersuchungen der Explosionskenndaten von Mischungen aus mehreren gas- und staubförmigen Brennstoffkomponenten und Luft, Fortschr.-Ber. VDI-Z., Reihe 3, Nr. 64, 1982
66. W. Berthold: Mindestzündenergie – Prüfverfahren VDI-Berichte 494, S. 105–108, 1984
67. BG-Chemie: Richtlinie Nr. 4: "Statische Elektrizität"
68. AK: "Brennbare Stäube": Sicherheitsmassnahmen gegen Staubbrände und Staubexplosionen, Ausgabe 1, Okt. 1973, Abschn. 2.1
69. W. Bartknecht: Untersuchungen über das Explosions- und Zündverhalten von Beschichtungspulvern, Ciba-Geigy AG, Zentraler Sicherheitsdienst, Fachgruppe Explosionstechnik, Bericht vom 19. 7. 1983 (unveröffentlicht)
70. J. P. Zeeuwen, G. F. M. van Laar: Ignition Sensitivity of Flammable Dust-Air Mixtures, Prins Maurits Laboratorium TNO (Holland), 1985
71. W. Bartknecht: Untersuchungen über das Explosions- und Zündverhalten von Flock, Ciba-Geigy AG, Zentraler Sicherheitsdienst, Fachgruppe Explosionstechnik, Bericht vom August 1983 (unveröffentlicht)
72. H. Franke: Bestimmung der Mindestzündenergie von Kohlenstaub/Methan/Luft-Gemischen (hybride Gemische), VDI-Berichte 304, S. 69, 1978
73. B. Maurer: Elektrostatische Entladungsvorgänge als Zündquellen, VDI-Berichte 494, S. 25–33, 1984
75. M. Steinicke: Gasbewegung und Turbulenz bei Explosionen in einer langgestreckten Bombe, Dissertation TH, Braunschweig, 1943
76. E. Schmidt, H. Steinecke, U. Neubert: Aufnahmen der Verbrennung von Gasgemischen in Rohren mit dem Eigenlicht der Flamme bei Schlierenbeleuchtung, VDI-Forschungsheft 431, 1951
77. G. Dammköhler: Gasbewegung in einem geschlossenen Verbrennungsraum bei einseitiger Zündung, Jahrbuch der deutschen Luftfahrtforschung, 1938
78. W. Bartknecht: Explosionstechnische Kennzahlen brennbarer Stäube in Rohren mit en-

gen Querschnitten, Internationales Symposium für Staubexplosionsgefahr im Bergbau und Industrie, Karlsbad (CSSR), 1972
79. H. R. Hupe: Staubexplosionen und Sicherheitsmassnahmen: "Untersuchungen im praxisnahen Maßstab bei spezieller Berücksichtigung von Gummi-, Holz- und Lackschleifstäuben", Dissertation TH Aachen, 1957
80. Berufsgenossenschaft der chemischen Industrie: Richtlinien zur Vermeidung der Gefahren durch explosionsfähige Atmosphäre mit Beispielsammlung: – Explosionsschutz-Richtlinien – Richtlinie Nr. 11, 1985, Druckerei Winter, 6900 Heidelberg
81. K. Isselhard: Die Aussagen der "Richtlinien für die Vermeidung von Gefahren durch explosionsfähige Atmosphäre mit Beispielsammlung" zur Vermeidung von Staubexplosionen, VDI-Berichte 304, S. 13–16, 1978
82. F. Schmalz: Untersuchungen über die Explosionsfortpflanzung aus einem druckentlasteten Behälter bei pneumatischer Förderung von Polyaethylengriess, Ciba-Geigy AG, Zentraler Sicherheitsdienst, Fachgruppe Explosionstechnik, Bericht D 17/82 vom 1.4.1982 (unveröffentlicht)
83. F. Lai: Explosibility of Grain Dust Collected from Grain Treated with Oil as a Dust Suppressant, Control of the Risks in Handling and Storage in Granular Foods, Paris, April 1985
85. W. Wiemann: Einfluss der Temperatur auf die Sauerstoff-Grenzkonzentration bei der Inertisierung, 9. Internationales Kolloquium für die Verhütung von Arbeitsunfällen und Berufskrankheiten in der chemischen Industrie, Luzern, Juni 1984
85. IVSS: Staubexplosionsgefährdete Maschinen und Apparate: "Vorbeugende und konstruktive Schutzmassnahmen", Arbeitskreis 6: "Staubexplosionen", 2/1987
86. W. Bartknecht: Untersuchung über das Zündverhalten von mechanischen Funken in Staub/Luft-Gemischen, Jahresbericht 1985 zum BMFT/HdA-Forschungsvorhaben: 01 VD 0220, Ciba-Geigy AG, Zentraler Sicherheitsdienst, Fachgruppe Explosionstechnik, (unveröffentlicht)
87. C. W. Wegst: Stahlschlüssel, Verlag Stahlschlüssel Wegst GmbH, D-7142 Marbach, 1986
88. K. Ritter: Zündwirksamkeit mechanisch erzeugter Funken gegenüber Gas/Luft- und Staub/Luft-Gemischen, Dissertation Universität Karlsruhe (TH), 1984
89. Berufsgenossenschaft der chemischen Industrie: Richtlinie: "Statische Elektrizität". Ausgabe 4/1980
90. G. Lüttgens, P. Boschung: Elektrostatische Aufladung: Ursachen und Beseitigung, Export-Verlag, D-7031 Grafenau, 1980
91. B. Maurer: Elektrostatische Entladungsvorgänge als Zündquellen, VDI-Berichte 494, Seite 119–127, 1984 und 9. Internationales Kolloquium für die Verhütung von Arbeitsunfällen und Berufskrankheiten in der chemischen Industrie, Luzern, 1984, S. 160–184
92. M. Glor: Hazards due to Electrostatic Charging of Powders, Journal of Electrostatics, 16, S. 175–191, 1985
93. G. Schuber: Einfluss des Vordrucks auf die Sauerstoff-Grenzkonzentration von Stäuben, Ciba-Geigy AG, Zentraler Sicherheitsdienst, Fachgruppe Explosionstechnik, Juni 1984 (unveröffentlicht)
94. C. Donat: Apparatefestigkeit bei Beanspruchung durch Staubexplosionen, VDI-Berichte 304, S. 139–149, 1978
95. C. Donat: Explosionsfeste Bauweise von Apparaturen, VDI-Berichte 494, S. 161–167, 1984
96. C. Donat: Explosionsfeste Bauweise, 9. Internationales Kolloquium für die Verhütung von Arbeitsunfällen und Berufskrankheiten in der chemischen Industrie, Luzern 1984, S. 665–991
97. G. N. Kirby, R. Siwek: Preventing Failures of Equipment Subject to Explosions, Chemical Engineering, June 23, 1986
98. W. Grein, H. O. Braubach, D. Wiesner: Technische Sicherheit und Verfügbarkeit von Chemieanlagen, Chem.-Ing.-Tech. 48. Jahrgang, Nr. 4, 1976
99. A. Schaerli: Chemische Industrie – Ein gefährlicher Nachbar? Risiken, Sicherheitsmassnahmen, Vortrag vor dem Rotary-Club, Basel, März 1977
100. (anonym): Verordnung über Druckbehälter, Druckgasbehälter und Füllanlagen, Carl Heymann-Verlag KG, Köln

101. VdTÜV: AD-Merkblätter
102. Arbeitskreis "Brennbare Stäube" der chemischen Industrie: Richtlinie "Explosionsdruckstossfeste Behälter und Apparate", Entwurf Juli 1985 (in Vorbereitung)
103. Basler chemische Industrie: Ex-Druckstossfeste Behälter und Apparate für brennbare Flüssigkeiten und Stäube: Berechung, Bau, Prüfung, Basler Norm 98, Entwurf Juli 1985 (in Vorbereitung)
104. W. Bartknecht: Druckentlastung von Staubexplosionen in Grossbehältern, Schriftenreihe: "Humanisierung des Arbeitslebens", Band 78, Herausgeber: Der Bundesminister für Forschung und Technologie, VDI-Verlag, Düsseldorf 1986
105. A. Aelling, R. Gramlich: Einfluss von Ausblasrohren auf die Explosionsdruckentlastung, VDI-Berichte 494, S. 175–183, 1984
106. R. Siwek: Additional Effects to take into Account in order to get Effective Explosion Venting, Europex-Seminar-Course, B-2540 Hove-Antwerpen, 1986
107. K. Ritter: Beispiele des Anlagenschutzes mit Kostenbetrachtung, VDI-Berichte 304, S. 159, 1978
108. M. Y. Brunner: Bauwerksbeanspruchungen durch die Rückstosskräfte druckentlasteter Staubexplosionen in Behältern, Diss. ETH Zürich Nr. 7223, 1983, VDI-Berichte 494, S. 227–232, 1984
109. M. Hattwig, M. Faber: Rückstosskräfte bei Explosionsdruckentlastung, VDI-Berichte 494, S. 219–226, 1984
110. M. Faber: Rückstosskräfte bei Explosionsdruckentlastung, 9. Internationales Kolloquium für die Verhütung von Arbeitsunfällen und Berufskrankheiten in der chemischen Industrie, Luzern 1984, S. 729–764
111. W. Bartknecht: Prüfung eines Wirbelschicht-Sprühgranulators bzw. -Trockners des Typs: S-8/T-8 unter Staubexplosionsbedingungen, Ciba-Geigy AG, Zentraler Sicherheitsdienst, Fachgruppe Explosionstechnik, Bericht D 28/83 vom Oktober 1983
112. W. Bartknecht: Explosions- und Funktionsprüfung an Wirbelschichttrocknern, Ciba-Geigy AG, Zentraler Sicherheitsdienst, Fachgruppe Explosionstechnik, Bericht D 2/85 vom Februar 1985
113. Hauptverband der gewerblichen Berufsgenossenschaften: Sicherheitsregeln für den Explosionsschutz bei der Konstruktion und Errichtung von Wirbelschicht-Sprüh-Granulatoren, Wirbelschichttrocknern und Wirbelschicht-Coatinganlagen, Ausgabe 10, 1980, Carl Heymanns KG, Köln
114. W. Bartknecht: Staubexplosionsversuche in einer Kohlenmahlanlage, Ciba-Geigy AG, Zentraler Sicherheitsdienst, Fachgruppe Explosionstechnik, Wochenrapport 1/80 (unveröffentlicht)
115. W. Bartknecht: Staubexplosionsversuche in einer über den Zuteiler entlasteten Kohlenmahlanlage, Wochenrapport 9/80 (unveröffentlicht)
116. S. Radandt: Staubexplosionen in Silos, Symposien: Heft 9,12, 14, 1981–1985, Berufsgenossenschaft Nahrungsmittel und Gaststätten, Knopf-Druck, 6803 Edingen-Neckarhausen
117. S. Radandt: Explosionsdruckentlastung in langgestreckten runden Silos, VDI-Berichte 494, S. 185–197, 1984
118. S. Radandt: Explosionsdruckentlastung langgestreckter Silos, 9. Internationales Kolloquium für die Verhütung von Arbeitsunfällen und Berufskrankheiten in der chemischen Industrie, Luzern, 1984, S. 697–722
119. W. Bartknecht: Effectiveness of Explosion Venting as a Protective Measure for Silos, Plant/Operations Progress (Vol. 4, No 11), January 1985
120. W. Bartknecht, R. E. Bruderer: Explosionsablauf von Maisstärke in einem druckentlasteten 20 m^3-Silo, Ciba-Geigy AG, Zentraler Sicherheitsdienst, Fachgruppe Explosionstechnik, Bericht vom 22. Mai 1985
121. W. Bartknecht: Prüfung von Bauelementen für einen Bandförderer auf Druckstossfestigkeit, Ciba-Geigy AG, Zentraler Sicherheitsdienst, Fachgruppe Explosionstechnik, Berichte D 30/82 und D 26/83 (unveröffentlicht)
122. W. Bartknecht: Die Schutzmassnahme, Explosionsunterdrückung, Sonderausgabe des "Ladenburger Kreises", März 1977, Herausgeber: Total-Foerstner & Co. AG, D-6802 Ladenburg

123. DEUGRA: Explosionsunterdrückung in Industriebetrieben, Gesellschaft für Brandschutzsysteme mbH, D-4030 Ratingen, 1985
124. D. Schneider: Erfahrungen mit Staubunterdrückungssystemen in Mahlanlagen, VDI-Berichte 494, S. 293–307, 1984
125. D. Schneider: Erfahrungen mit Schutzmassnahmen gegen Staubexplosionen, 9. Internationales Kolloquium für die Verhütung von Arbeitsunfällen und Berufskrankheiten in der chemischen Industrie, Luzern, 1984, S. 489–534
126. International Standard ISO 6184/4: Explosion Protection Systems – Part 4: Determination of Efficacy of Explosion Suppression Systems, International Organization for Standardization, 1985
127. P. Moore, W. Bartknecht: Suppression of Gas and Dust Explosions in Large Volumes, 5th International Symposium Loss Prevention and Safety Promotion in the Process Industries, Cannes/France, September 1986
128. W. Czajor: Explosionstechnische Entkoppelung von Apparaturen, VDI-Berichte, 494, S. 233–238
129. E. W. Scholl: Ein Verfahren zum Löschen von Schlagwetter-, Kohlenstaub- und Methan-Kohlenstaub-Explosionen, Glückauf-Forschungshefte, 29. Jahrgang, Heft 4, 1968
130. W. Bartknecht: Löschmittelsperre, Sonderausgabe des "Ladenburger-Kreises", Oktober 1977, Herausgeber: Total-Foerstner & Co. AG, D-6802 Ladenburg
131. W. Bartknecht: Staubexplosionsversuche in einem verstärkten Gurtelevator der Fa. Bühler-MIAG GmbH (Braunschweig), Ciba-Geigy AG, Zentraler Sicherheitsdienst, Fachgruppe Explosionstechnik, 1981
132. F. Kossebau: Explosionsgeschützte Elevatoren, Bühler-MIAG-Nachrichten, 219, S. 20–21, 1981
133. G. Schuber: Zünddurchschlagverhalten von Staub/Luft-Gemischen und von hybriden Gemischen, Jahresbericht 1984/1985 zum BMFT/HdA-Forschungsvorhaben, Ciba-Geigy AG, Zentraler Sicherheitsdienst, Fachgruppe Explosionstechnik (Veröffentlichung in Vorbereitung)
134. IRS: Prospektblatt: Schnellschuss-Schieber, Industrie-Rationalisierungssysteme, D-6100 Darmstadt, 1985
135. SISTAG: Prospektblatt: WEY-Explosionsschutz- Schieber, Typ SLC-Ex, Sidler Stalder AG, Maschinenfabrik, CH-6274 Eschenbach
136. H. Stalder: Explosionsschutz-Schieber: Massnahmen gegen die Auswirkungen von Explosionen und Detonationen, 10. Internationales Kolloquium für die Verhütung von Arbeitsunfällen und Berufskrankheiten in der chemischen Industrie, Frankfurt 1985, S. 525–555
137. Silo-Thorwesten: Prospektblatt: Druck-Entlastungsschlot, Silo-Thorwesten GmbH, D-4720 Beckum, 1985
138. S. Radandt: Bestimmung der Explosionsdruck-Entlastungsflächen in Übereinstimmung mit den am häufigsten angewendeten Richtlinien innerhalb und außerhalb Europas, Europex-Seminar: "Druckentlastung von Staubexplosionen in Behältern", 1986
139. W. Bartknecht: Druckentlastung von Staubexplosionen in Behältern. Staub-Reinhaltung in der Luft, Nr. 9, September 1986, S. 368–373
140. W. Bartknecht: Massnahmen gegen gefährliche Auswirkungen von Staubexplosionen in Silos und Behältern, Jahresbericht 1985 zum BMF/HdA-Forschungsvorhaben vom Februar 1986 (unveröffentlicht)

10 Symbols and Abbreviations

a		factor for sizing vent
A		vent area
abs		absolute
AD		Arbeitsgemeinschaft Druckbehälter
Al		aluminum
b		factor for sizing
BAM		Bundesanstalt für Materialprüfung
c		factor for sizing
C		capacitance
CL		class
Cr		chromium
d		main gap distance
D		diameter
D		Dernier
DIN		Deutsches Institut für Normung
DN		nominal diameter
dp/dt		rate of pressure rise
$(dp/dt)_{max}$		maximum rate of pressure rise
$(dp/dt)_{red}$		reduced rate of pressure rise
$(dp/dt)_{red,max}$		reduced maximum rate of pressure rise
D_R		diameter of the rubbed spot
E		energy
EE		electrical equivalent energy
e. g.		exempli gratia (for example)
Ex-RL		explosion protection guideline
f		function
F_R		recoil force
FRG		Federal Republic of Germany
g		grams
H		height
H		product humidity
h, hrs		hour(s)
Halon		halogenated hydrocarbons
HRD		high rate discharge
I		current
ID		inside diameter
IE		ignition energy

IEC	International Electrotechnical Commission
IP	probability of ignition
i.e.	id est (that is)
J	joule
kJ	kilojoules
K	degrees Kelvin (Celsius absolute)
K_G	constant for gases
K_{St}	constant for dusts
kg	kilograms
k_1	relay
l	length
l	liter
L	inductance
LEL	lower explosible limit
LMIE	lowest minimum ignition energy
LOC	limiting oxygen concentration
lg	gap length
log	common logarithm of
lp	length of pipe
m	meters
M	median value
MESG	maximum experimental safe gap
MIE	minimum ignition energy
ML	material-to-air loading
max	maximum
mm	millimeters
ms	milliseconds
N	number of extinguishers
N	newton
Nm	newton meters
N_2	nitrogen
OC	oxygen concentration
O_2	oxygen
p	pressure
p	pneumatic
p_a	activation pressure (of suppression systsems)
p_A	thrust
p_{atm}	atmospheric pressure
PE	polyethylene
p_{ex}	explosion pressure
PN	nominal pressure
p_{N2}	pressure of the driving medium N_2
PVC	polyvinyl chloride
p_i	initial pressure
p_{max}	maximum explosion pressure
p_{red}	reduced explosion pressure
$p_{red,max}$	reduced maximum explosion pressure
Q	volume of air

R	resistor
RL	guideline
s	seconds
SA	surface area
sect.	Section
St	dust explosion class
t	time
T	temperature
T_A	autoignition temperature
t_d	ignition delay time
t_g	grinding time
T_{Gl}	glowing wire temperature
TI	ignition temperature
t_i	induction time
t_o	oven temperature
T_{opt}	optimum temperature
t_r	rubbing time
t_{red}	reduced time
t_1	duration of combustion
t_{1min}	minimum duration of combustion
t_2	induction time
t_{2min}	minimum induction time
U	voltage
UEL	upper explosible limit
v	velocity
V	volume of vessel
VDE	Verband Deutscher Elektrotechniker
VDI	Verein Deutscher Ingenieure (German Society of Engineers)
v_{ex}	flame velocity
$v_{ex,max}$; v_{max}	maximal flame velocity
vol	volume
vr	speed of rotation
vs	versus
wt	weight
α	dynamic coefficient
ΔT	temperature difference
μm	micron; 10^{-6} m
Ω	ohm
%	percent
°C	degrees Celsius

11 Conversion Factors

1 mm	= 0.0394″
1 cm	= 0.3937″
1 m	= 3.2808′
1 km	= 0.6214 st.mi
1 inch	= 2.5400 cm
1 foot	= 0.3048 m
1 status mile	= 1.6093 km
1 cm^2	= 0.1550 sq.in
1 m^2	= 10.7639 sq.ft
1 m^2	= 1.1960 sq.yd
1 m^2	= 15.499 sq.in
1 square inch	= 6.4516 cm^2
1 square foot	= 0.0929 m^2
1 square yard	= 0.8361 m^2
1 square inch	= 6.452 · 10^{-4} m^2
1 cm^3	= 0.06102 cu.in
1 l	= 0.03531 cu.ft
1 m^3	= 35.31 cu.ft
1 m^3	= 1.308 cu.yd
1 cubic inch	= 16.387 cm^3
1 cubic foot	= 28.314 l
1 cubic foot	= 0.02832 m^3
1 cubic yard	= 0.7646 m^3
1 l	= 0.2642 (USA) gallon
1 l	= 0.220 (Brit) gallon
1 (USA) gallon	= 3.7854 l
1 (Brit) gallon	= 4.5461 l
1 g	= 15.4323 grain
1 kg	= 2.2046 pounds
1 grain	= 0.0648 g
1 pound	= 0.4536 kg
1 bar	= 14.504 psi
1 bar	= 10^5 Pa
1 Pa	= 0.000145 psi

1 pound-force/inch^2 = 0.06895 bar
1 pascal = 0.00001 bar
1 pound-force/inch^2 = 6894.8 Pa

x°C = 1.8x + 32°F

x°F = 5/9 (x-32)°C

12 Subject Index

A
Autocatalytic decomposition 42, 43
Autoignition 36
–, induction time 39
–, – –, dependency of storage temperature and test volume 39
–, heat balance 36
–, heat generation 36
–, heat transmission 37
–, storage temperature 36
–, surrounding temperature 36
–, temperature 36
– –, determination 37
– –, method dependency 36
– –, volume dependency 36, 37
– –, –, EC-method 37
– –, – as per Grever 37
– –, –, hot storage test in a wire mesh basket 39

B
Blow off pipes, see relief pipes
Brush discharge 157
Burning behavior 28, 29
– –, burn off velocity 29, 31
– –, – – –, determination 31
Burning rate, see Burning behavior

C
Calculation model, explosion pressure venting
– –, silos 251
– –, vessels 248
Carbonization gases, see decomposition gases
Chemical ignitors, see pyrotechnical ignitors
Coal pulverizer 182
Combustibility test 29
– – at elevated temperature 30
– – at reduced pressure 32
– – at room temperature 29
– –, appearance 29, 30
– –, Combustion class 29, 30
– –, – –, temperature dependency 30
– –, under vacuum 32

Combustible dusts 51
– –, explosible limits 54
– –, explosion indices 56–80
– –, ignition temperature 115–119
– –, minimum ignition energy 96–109
Combustion time, see duration of combustion
Conical pile discharge 158, 159
Constructive explosion protection 161
– – –, design for the maximum explosion pressure 163
– – –, diverting 215
– – –, explosion pressure venting 166
– – –, explosion suppression 203
Cubic Law 61
– –, validity 61, 64, 65

D
Decomposition energy 44, 45
Decomposition gases 32, 40, 43, 45, 50
Decomposition temperature, see exothermic decomposition
Deflagration 32
–, ashes 32
–, –, combustibility 32
–, avoidance 32
–, decomposition gases 32
–, –, explosion hazard 32
–, –, pressure build-up 32
–, initiated 32
–, stoppage 32
–, temperature 32
–, test 32, 33
–, –, laboratory test 33
–, –, –, laboratory test at elevated temperature 33
–, –, laboratory test at room temperature 33
–, –, laboratory test inerting 33
–, –, screening test 32
–, –, screening test, test under vacuum 32
–, –, screening test, Witt scher pot 33
–, velocity 33
Dependency on volume, see volume dependency
Detector for explosion suppression 205

Subject Index

Diversion 215
–, diverter 243
–, extinguishing barrier 215
–, rapid action valve: butterfly type 241
–, rapid action valve: float type 241
–, rapid action valve: gate typ 232
–, rotary air locks 229
Diverter 243
Duration of combustion 107, 108
Dust
–, airborne dust 51
–, definition 25
–, dust layers 28
–, settling velocity 25, 26
Dust accumulation
– –, acceptable 12
– –, –, cubically-shaped 13
– –, –, not cubically-shaped 13
Dust explosions
– –, bakeries 10
– –, coal dust 4
– –, combines 3
– –, frequency 4, 7
– –, –, types of equipment 7
– –, –, dust types 7
– –, –, types of ignition sources 8
– –, grain/flour 2, 5
– –, mechanics 14
– –, number 2, 4
– –, silos 4, 7
– –, sugar 4, 10, 11
Dust explosion class 78
– – –, frequency 79
– – –, increasing 90
Dust investigations 23
– –, apparatus 15, 56, 62
– –, dispersal procedure 22, 23, 56, 62
Dust layers 28
Dust temperature 28, 30, 31, 33

E

Electrical equivalent energy 148
Electrostatic ignition sources 156
– – –, brush dischargers 157
– – –, conical pile dischargers 158
– – –, lightning-like discharges 158
– – –, propagating brush discharges 158
– – –, spark discharges 156
Elevator 223
–, efficacy of extinguishing barrier 225, 226
Endothermic reaction 40
Exothermic decomposition 40
– –, autocatalytic 42, 43
– –, decomposition energy 44, 45
– –, decomposition gases 40
– –, – –, combustibility 43

– –, – –, pressure build-up 40
– –, – –, explosion hazard 40, 44
– –, – –, type of test 40
– –, – –, quantity 43
– –, decomposition temperature 40
– –, – –, dependency on determination method 40
– –, heat accumulation 40
– –, – –, adiabatic condition 40, 45
– –, – –, critical temperature 40, 45
– –, – –, heat elimination 44
– –, – –, heat elimination, capacity of heat elimination 44
– –, – –, heating efficiency 44
– –, heat production 44
– –, – –, rate of heat production 44
– –, self-heating 40, 46
– –, –, induction time 45, 46
– –, thermal explosion 40, 44, 45
– –, type of test
– –, – – –, differential thermal analysis 44
– –, – – –, DTA, see differential thermal analysis
– –, – – –, as per Lütolf 40
– –, – – –, with nitrogen, Grever-Oven 44
– –, – – –, hot storage test in the Dewar-tube 45
– –, –, hot storage test under pressure build-up 46
– –, –, hot storage test with temperature programmed conditions 46
Explosibility 11, 27, 47
–, definition 53
–, friction sensitivity 47
–, – –, test type 49
–, – –, test type, apparatus as per BAM 49
–, impact sensitivity 48
–, – –, test type 48
–, – –, test type, drop weight as per Koenen 48
–, – –, test type, drop weight as per Lütolf 48
–, thermal sensitivity 50
–, – –, test type 50
–, – –, test type, steel tube test 50
–, – –, test type, steel tube test, limiting diameter 51
Explosion characteristics, see explosion indices
Explosion doors, see explosion flaps
Explosion flaps
– –, bucket elevator 201
– –, – –, influence of weight of the flaps 201
– –, pipes 203
– –, silos 194
– –, –, venting efficacy 196

– –, vessels 174
– –, –, venting efficacy 175
Explosion hazard, see explosibility
Explosion indices
– –, combustible dusts 56–80
– –, flock 84–86
– –, hybrid mixtures 88–91
Explosion limits 54
– –, combustible dusts 54
– –, – –, influence of temperature 56
– –, – –, influence of ignition energy 55
– –, – –, lower 54
– –, – –, upper 54
– –, flock 82
– –, –, influence of ignition energy 83
– –, hybrid mixtures 87
– –, – –, influence of flammable gases 87
– –, – –, influence of type of flammable gases 88
Explosion pressure 56
– –, combustible dusts 56
– –, – –, influence of humidity 72
– –, – –, influence of filling ratio 69
– –, – –, influence of ignition energy 77
– –, – –, influence of initial pressure 73
– –, – –, influence of particle size 69
– –, – –, influence of solvents 72
– –, – –, influence of temperature 73
– –, – –, influence of turbulence 58
– –, – –, influence of volume 61
– –, – –, influence of water 71
– –, flock 84
– –, –, influence of fiber 84
– –, –, influence of ignition energy 86
– –, frequency 79
– –, hybrid mixtures 88
– –, – –, influence of flammable gases 89
– –, – –, influence of type of flammable gases 91
– –, optimum dust concentration 91
Explosion pressure venting 166
– – –, belt elevator 201
– – –, coal pulverizer 181
– – –, fluidized bed driers 177
– – –, – – –, explosion pressure, reduced 167
– – –, – – –, explosion pressure, maximum 168
– – –, large vessels 183
– – –, pipelines 200
– – –, –, detonation 239
– – –, –, explosion 200
– – –, –, silos 187
– – –, –, explosion velocity 192
– – –, –, influence of arrangement 190
– – –, –, influence of partial filling 191
– – –, –, pneumatic conveying 194, 249–253
– – –, –, pneumatic conveying, pressure-area curves 252
– – –, –, pneumatic conveying, influence of bulk density 250
– – –, –, pneumatic conveying, calculation model 251
– – –, –, pressure rise, rate of 192
– – –, –, pressure venting installations 194
– – –, –, pressure venting installations, venting efficacy 196
– – –, vessels 166
– – –, –, explosion flaps 174
– – –, –, explosion flaps, venting efficacy 175
– – –, –, explosion pressure, reduced 167
– – –, –, explosion pressure, maximum 167
– – –, –, influence of relief pipes 172, 173
– – –, –, large vessels 183
– – –, –, large vessels, nomograms 249
– – –, –, large vessels, calculation model 248
– – –, –, nomograms 171
– – –, –, pressure rise, reduced rate of 167
– – –, –, pressure rise, maximum 167
– – –, –, recoil forces 175
Explosion protection guideline 126
Explosion resistant design 163
– – –, explosion pressure resistant 161–163
– – –, explosion pressure shock resistant 161, 162, 165
Explosion suppression 203
– –, activation pressure 209
– –, detectors, see sensors
– –, dispersion of extinguishing agents 206, 207
– –, extinguishers 206, 207, 212
– –, extinguishing agent requirement 210, 211, 214
– –, influence of extinguishing agents 209
– –, limits of application 211
– –, sensors 204
– –, telescopic hemispherical nozzle 208
Extinguishers
–, for explosion suppression 206, 207, 212
–, for extinguishing barriers 217
Extinguishing agent
– –, for explosion suppression 209
– –, for extinguishing barriers 217
Extinguishing agent requirement
– – –, for explosion suppression 210–214
– – –, for extinguishing barriers 219
Extinguishing barriers 215
– –, effectiveness 216
– –, –, elevator 223
– –, –, pipelines 220

Subject Index

Extinguishing barriers, effectiveness, silos 226
– –, extinguishing agent requirement 210–214, 218, 219

F

Fan-shaped nozzle 218
Fiber 26
–, flock 26
Flame jet ignition 215
Flammability 28
–, test 28
–, – at elevated temperature 28
–, – at room temperature 28
–, –, ignition sources 28
–, –, lighter with flintstone 28
–, –, lighter with flintstone, sparks 28
–, –, lighter with flintstone, gas flames 28
–, –, lighter with flintstone, mechanical sparks 28
–, –, lighter with flintstone, matches 28
–, of deposited combustible dusts 28
Flintstone friction sparks 147
Flock 81
–, explosion indices 84–86
–, explosion limits 82, 83
–, minimum ignition energy 109–112
Fluidized bed driers 177–180
Foreign particle 32, 48
Friction sensitivity, see explosibility
Friction sparks 150

G

Gap, maximum experimental safe 230
Glowing temperature 154
Glowing wire coil 117
– – –, comparison with minimum ignition energy 118
– – –, ignitability 118
Grewer Th. 37, 38

H

Hazard triangle 127
Heat accumulation, see exothermic decomposition
Historical review 2
Hot surfaces, see glowing temperature
Hybrid mixtures 86
– –, explosion indices 88–91
– –, explosion limits 87, 88
– –, minimum ignition energy 112–114

I

Ignitability, see Flammability 28
Ignition sources 19
– –, chemical, pyrotechnical 23
Ignition temperature 115
– –, apparatus for determination 116

– –, – – –, BAM procedure 116
– –, – – –, Godbert-Greenwald procedure 116
Impact sensitivity, see explosibility
Impact sparks 151
Induction time 107, 108
Inerting 129
–, nitrogen 129
–, –, limiting oxygen concentration 130
–, –, limiting oxygen concentration, combustible dusts 130
–, –, limiting oxygen concentration, combustible dusts, influence of volume of apparatus 132, 160
–, –, limiting oxygen concentration, combustible dusts, influence of temperature 134
–, –, limiting oxygen concentration, combustible dusts, correlation K_{St}-value 133
–, –, limiting oxygen concentration, combustible dusts, correlation dust concentration 134
–, –, limiting oxygen concentration, combustible dusts, correlation ignition temperature 133
–, –, limiting oxygen concentration, hybrid mixtures 137
–, –, limiting oxygen concentration, hybrid mixtures, influence of flammable gases 137–140
–, –, limiting oxygen concentration, hybrid mixtures, influence of oxygen concentration 138
–, –, maximum allowable oxygen concentration 129, 135
–, solids 141
–, –, correlation dust concentration 142
–, –, correlation minimum ignition energy 141
–, –, influence of ignition energy 143
–, –, limiting inert dust concentration 141
–, vacuum 140
Isothermal conditions 37, 39, 42, 44

K

Koenen H. 48
K_{St}-value 61
–, frequency 79

L

Laboratory apparatus 62, 63, 66, 67
Large-scale vessels
– –, explosion pressure venting 188
– –, explosion suppression 213, 214
Le Chatelier Law 87
Lightning-like discharge 158, 159

Subject Index

Limiting inert dust concentration 141
Limiting oxygen concentration 129
---, combustible dusts 130
---, hybrid mixtures 137
Localization of heat, see exothermic decomposition Lütolf J. 40, 48

M

Material influence, see exothermic decomposition
Material safety specification 27
---, airborne dust, see dust clouds
---, dust clouds 51
---, dust layers 28
Mechanical sensibility, see explosibility
Mechanically generated sparks 145
Median value 52, 69
Minimum ignition energy, lowest 93
---, apparatus for determination 93
---, combustible dusts 93
---, --, coating powders 105
---, --, coating powders, influence of added aluminum 105
---, --, coating powders, influence of dust concentration 105
---, --, correlation duration of combustion 108
---, --, correlation induction time 108
---, --, frequency 102
---, --, ignition probability 100, 101
---, --, influence of vessel volume 97, 98
---, --, influence of dust concentration 97
---, --, influence of gap distance 99
---, --, influence of humidity 106
---, --, influence of ignition delay time 97
---, --, influence of inductivity 100
---, --, influence of initial pressure 103
---, --, influence of particle size 104
---, --, influence of temperature 107
---, --, statistically transitory region 101

O

Optical flame sensors 217
Oxidation reaction, see autoignition
Oxygen concentration, maximum allowable 129

P

Partial filling 68
Particle size 11, 24, 34–36
--, median value 52, 53, 69
Particle size distribution 52, 53
---, median value 52, 53
Pipelines

-, course of explosion 119
-, explosion pressure 120
-, --, maximum 121
-, explosion velocity 120
-, -, maximum 121, 123
-, explosion indices 123, 124
Pneumatic conveying, silos 194, 248–252
Powder, see dust
Pressure relief arrangements
---, silos 194
---, -, relief efficiency 196
---, vessels 174
---, -, relief efficiency 175
Pressure rise
--, combustible dusts 57
--, --, influence of humidity 72
--, --, influence of ignition energy 77, 78
--, --, influence of initial pressure 73
--, --, influence of partial filling 68
--, --, influence of particle size 69
--, --, influence of solvents 72
--, --, influence of temperature 75
--, --, influence of turbulence 59
--, --, influence of volume 61
--, --, influence of water 72
--, flock
--, -, influence of fiber 84
--, -, influence of ignition energy 86
--, hybrid mixtures 88
--, --, influence of combustible gases 89
--, --, influence of type of combustible gases 91
--, maximum 57
--, optimum dust concentration 91
--, rat of 57
Pressure rise, reduced rate of 167
--, ---, maximum 167
Preventing the formation of an explosible dust/air mixture 128
Prevention of effective ignition sources 144
-----, hot surfaces 154
-----, mechanically generated sparks 145
-----, ---, boundaries 146
-----, ---, electrical equivalent energy 146
-----, ---, flintstone friction sparks 147
-----, ---, friction sparks 151
-----, ---, impact sparks 151, 152
-----, ---, steel sparks 151, 152, 153
-----, static electricity 156
-----, --, brush discharges 157
-----, --, conical pile discharges 158
-----, --, lightning-like discharges 158

Subject Index

Prevention of effective ignition
  sources, static electricity,
  propagating brush discharges 158
– – – – –, – –, spark discharges 156
Preventive explosion protection 127
– – –, inerting 129
– – –, preventing the formation of an
  explosible dust/air mixture 129
– – –, prevention of effective ignition
  sources 146
Propagating brush discharge 158
Pyrotechnical ignitors 23

R
Radiation 14
Rapid action valve
– – –, butterfly type 241
– – –, float type 241
– – –, – –, activated with a device 243
– – –, – –, explosion pressure activated 242
– – –, gate type 232
– – –, – –, venting elements for 238
Recoil forces, explosion pressure
  venting 175
Relief pipes for pressure venting 172
– – – – –, nomogram 173
Resistance of apparatus 162
Rotary air locks, rotary locks 229

S
Silo
–, course of explosion 191, 192

–, efficiency of extinguishing barrier 193
–, explosion pressure venting 187, 189, 193
Smolder temperature 34
– –, determination 35
– –, –, according to IEC 35
– –, influence of particle size 34, 35
– –, – of bulk density 34, 35
– –, – of thick layer 34, 35
Spark discharge 157
Standard deviation 66
Static electricity 156
Steel sparks 151, 152
Storage conditions 36, 39, 44
Storage temperature 36, 39
Surface temperature 34
Surrounding temperature 36, 45

T
Telescopic hemispherical nozzle 208
Thermal explosion 40, 44, 45
Thermal sensibility, see explosibility
Thermal stability, see exothermic
  decomposition
Three-electrode arrangement 93
Time characteristics 108
Turbulence 51
–, flammable gases 51, 52

V
Volume dependecy 39, 40, 45